DESIGN FOR PRIVACY
KEEPING PERSONAL INFORMATION PRIVATE

Robert Stribley

NEW YORK   2025

"Privacy lies at the heart of the user-technology relationship. Robert Stribley offers invaluable tactical advice and broader context that designers need to make that relationship as safe as possible."

—Alexandra Schmidt
Author of *Deliberate Intervention*

"The stakes for data misuse or accidental exposure are incredibly high; this book is a must-read for any design leader looking to build products that respect and protect users' privacy."

—Bryan Hamilton
Global Head of Design, BNY

"Finally—a book that treats privacy not as a compliance box, but as a core design principle. Stribley's essential guide bridges the inexplicable gap between these disciplines."

—Pepe Borrás
Cofounder, Internet Freedom Festival,
Internet Freedom Expert and Product Director

"Privacy isn't about compliance: It's about helping people to negotiate their boundaries. Designers have long-needed expert advice on the arguments and tactics that make that principle a reality. At last, help has arrived."

—Cennydd Bowles
Technology Ethicist

"As a designer, you have more power over users' privacy than you may realize. Robert shows you how to leverage that power for good with practical tips on how to get your organization to care, plenty of real examples, and clear guidance. This is the go-to privacy book written by a designer, for designers."

—Heidi Trost
Author of *Human-Centered Security*

"The definitive privacy-by-design guide for these evolving early days of AI's design acceleration. It's essential reading to prevent catastrophic privacy failures that obliterate customer trust and your product success."

—Chad Borlase
SVP, North American Experience Design Lead,
Merkle, a dentsu company

Design for Privacy
Keeping Personal Information Private
By Robert Stribley

Rosenfeld Media, LLC

125 Maiden Lane

New York, New York 10038

USA

On the Web: www.rosenfeldmedia.com

Please send errata to: errata@rosenfeldmedia.com

Publisher: Louis Rosenfeld

Managing Editor: Marta Justak

Interior Layout: Danielle Foster

Cover Design: Heads of State

Indexer: Marilyn Augst

Proofreader: Sue Boshers

© 2025 Robert Stribley

All Rights Reserved

ISBN: 1-959029-66-5

ISBN 13: 978-1-959029-66-3

LCCN: 2025931267

Dedicated to my Dad, Robert Marshall Stribley

Thank you for being my role model for modesty, authenticity, commitment, responsibility, and pride in your work.
I just wish you were here to see this.

HOW TO USE THIS BOOK

Who Should Read This Book?

If you're engaged in the creation of digital products, then this book is for you. If you provide input *at all* into how online experiences are designed and deployed, then this book is for you. You may be a user experience designer, an interface designer, a product designer, a product manager, a strategist, a developer, a producer, or a project manager. Essentially, you are a designer. Additionally, anyone who has concerns around privacy online should find this a compelling and informative read. It's also my hope that many in the internet and human rights communities will embrace this book as a call to action to improve our lives online, too.

What's in This Book?

As our online lives continue to deepen, we find ourselves grappling with ever-evolving online privacy issues. Designers have a responsibility to address these issues and to consider how they can avoid them in their designs. This book will examine specific examples of privacy issues in digital design, as well as why it's increasingly important for designers to pay attention to them. The core of the book will focus on addressing these issues by considering the following key practices:

- Handle people's data responsibly.
- Avoid deceptive experience patterns.
- Advocate for better use of language.
- Provide tools that empower users to maintain their privacy.

You'll also learn how to build a culture that's friendly to the practice of privacy by design and also about some organizations that focus specifically on privacy. And you'll find some insight into how AI is creating new privacy issues and whether they can be addressed through design.

What Comes with This Book?

This book's companion website (https://rosenfeldmedia.com/books/design-for-privacy/) contains the book's diagrams and other illustrations—all available under a Creative Commons license (when possible) for you to download and include in your own presentations.

FREQUENTLY ASKED QUESTIONS

Why should I care about privacy? I have nothing to hide.

Like freedom of speech, a right to privacy is often something we may not give much consideration—until we need to. As you'll see, you never know what sudden changes to company policies or ownership or even regional or national politics and law will have upon your privacy. This book argues that in considering people's specific needs for privacy, we will inevitably create better experiences for us all. And, more broadly, we'll help to create a world where privacy is valued and treated as a right for us humans, not just a hurdle for companies and organizations to overcome. **Chapters 1** and **3** cover the ground that emphasizes most concretely why we should be concerned about privacy both as individuals *and* designers.

What motivation do companies have to resolve privacy issues if they're profiting from them?

Realistically, given the profit motive, some companies may not find ethical reasons for practicing privacy by design terribly compelling. However, such organizations should consider these additional reasons for maintaining privacy-respecting experiences, too: damage to their reputations, experience abandonment, loss of user base, and the penalties and fines that may be leveraged against them if they don't adhere to emerging privacy regulations. **Chapter 4** offers much more in the way of anecdotes and data to clarify the implications of these different factors.

Is privacy policy really doing anything to impact how companies approach privacy issues within their experiences?

As privacy policy evolves globally, it becomes increasingly more impactful. In Europe, for example, increasingly large fines are being leveraged against companies that don't adhere to privacy regulations. Further, regulations in many parts of the world place an increasing emphasis upon privacy as a right for individuals that outweighs the needs of business. Many countries, however, including the United States, still lag behind in this area. To co-opt a line from William Gibson, then, the future of privacy is here, but that future is not evenly distributed. **Chapter 11** covers this topic in detail.

How much help can designers really be with privacy issues?

It's true that some privacy issues are rooted in key business goals that designers play little or no role in formulating. It's also true that the language of privacy is almost always governed by a company's legal department. However, designers are still perfectly placed to argue and design for privacy-friendly solutions to business requirements, to imagine tools that help users maintain control of their personal information, and to introduce and frame privacy-oriented content, so that it's better understood by people accessing these experiences. **Chapter 3** is dedicated to explaining how well-situated designers really are to make a difference.

If you had to boil the resolution of privacy issues down to one word, what would it be?

"Consent." Most of the problems arising around privacy online come down to a lack of consent. People either lack the ability to consent how their data is used, or they're not clearly informed as to what they're consenting to. Further, they often have no way of fine-tuning their consent or withdrawing their consent if they decide they no longer wish to share their information. **Chapters 5, 6, 7,** and **8** each provide *a lot* of super specific guidance on how to properly inform people about what they're consenting to and how to design features which enable their consent transparently and honestly.

CONTENTS

How to Use This Book — vi
Frequently Asked Questions — viii
Foreword — xiv
Introduction — xvi

CHAPTER 1
A Deluge of Privacy Issues — 1
Issues with Data — 2
Invasive AI — 9
Cyberstalking and Bullying — 12
What Can Designers Address? — 13
The Takeaway — 14

CHAPTER 2
Defining Privacy — 15
What's the Difference Between Privacy and Security? — 18
Defining a Right to Privacy — 18
Some Relevant Privacy Terminology — 20
The Origins of Privacy by Design — 23
The Takeaway — 24

CHAPTER 3
Your Role as a Designer — 25
Privacy and the First Ethical Hackers — 26
History May Not Repeat but It Rhymes — 29
Why Me? — 31
Who Are These "Users" Anyway? — 33
The Takeaway — 35

CHAPTER 4
Why Should Business Care? — 37
Civic Responsibility — 38
Damage to Reputation — 39

Experience Abandonment	40
Loss of User Base	41
Penalties and Fines	44
The Takeaway	46

CHAPTER 5
Handle Data Responsibly — 47

Only Ask for the Data You *Really* Need	49
Always Obtain Explicit Consent	61
Maintain Data Transparency	63
Make Sure Users Can Delete Their Data	78
The Takeaway	86

CHAPTER 6
Avoid Deceptive Patterns — 87

Defining Deceptive Patterns	88
Deceptive Patterns Affecting Privacy	89
Approaches for Avoiding Deceptive Patterns	109
The Takeaway	118

CHAPTER 7
Use Language with Care — 119

Deciphering the Indecipherable	120
Aim for Clarity	123
Keep Content Honest	134
Make Navigating Content Easy	141
The Takeaway	146

CHAPTER 8
Provide Tools for Enabling Privacy — 147

Make Privacy Tools a Priority	148
Ensure That Privacy Features Are Easily Discoverable	154
Follow Best Practices for Privacy Features	161

Remind Users of Privacy Features 171
Never Change Privacy Settings Without
  Letting Users Know 174
The Takeaway 178

CHAPTER 9
Cultivating a Culture for Privacy by Design 179
Privacy by Design as a Practice 180
Practice Inclusive Design 182
Ensure Experiences Are Accessible 187
Reference Thoughtful Personas and Archetypes 188
Employ User Journeys and Stories 190
Evaluate Experiences for Harm 190
Drive Change as a Design Leader 198
Never Stop Learning 199
The Takeaway 200

CHAPTER 10
Exercises: Navigating Two Privacy Scenarios 201
Scenario 1: Friends, a Networking Site 203
Scenario 2: BookLuvr, a Book-Sharing App 204
The Takeaway 206

CHAPTER 11
The Evolving Impact of Privacy Policy 207
The Impact of Europe's GDPR 208
The California Consumer Privacy Act 211
The Rest of the United States 214
Evolving Global Efforts 215
Enforcing Penalties 217
Corporate Self-Regulation and Policy 218
Massive Companies Enforcing Privacy Unilaterally 219
The Takeaway 224

CHAPTER 12
AI and Privacy — 225
Lack of Transparency with Use of Data — 226
Accidental Exposure of Personal Data — 229
The Myth of Data Anonymization — 231
AIs Designing Deceptive Patterns — 232
AIs Listening in Everywhere — 233
Malicious Misuse of AI — 235
The Takeaway — 238

CHAPTER 13
Working on Privacy: Privacy as a Product — 239
Browsers — 240
VPNs — 243
Extensions — 244
Search — 252
Sites and Apps — 252
Projects — 253
The Takeaway — 254

Conclusion — 255
Getting Worse Before It Gets Better — 256
Demanding a Better Future — 257

Appendix A — 258

Appendix B — 260

Index — 261

Acknowledgments — 278

About the Author — 282

FOREWORD

"Are you a privacy expert? It's a simple yes or no question!"

When I'm under oath in a deposition for a lawsuit, this is the sort of thing an angry lawyer might shout at me. Sometimes they try to simplify complex issues like this to their advantage, so they can persuade the judge to disregard the insights I present on manipulative or deceptive design.

Of course, it's a flawed question—privacy expertise isn't a binary skill that you either have or don't. Privacy is a vast, multifaceted domain with diverse kinds of expertise, and it can be viewed through multiple lenses like legal, moral, technological, cultural—or the angle that Robert Stribley takes in this essential book: a design lens.

Design is a particularly important lens because it shows us how abstract business policy gets rendered into reality. It's how businesses tell users what they're going to do with their data and how businesses present choices to users. Since we're all human, we have natural limitations in our time, attention, and cognition, which makes us vulnerable. This means we can either be carefully supported or ruthlessly exploited by a business.

It's funny—when a business sends your private data somewhere, you usually don't find out about it. It's not as if you get a little electric shock from your keyboard each time your data gets sold to a third party. In fact, your immediate experience often gets a bit better when you choose low privacy settings. You get more relevant recommendations, nicer ads, and platforms usually stop nagging you about your privacy choices. Giving in and sharing everything you're asked for is often the easiest way to have a pleasant online experience. The consequences of our privacy decisions can be hidden and out of reach, which means it's difficult for us to grasp the full impact of our actions—particularly when this involves difficult concepts such as algorithmic discrimination or erosion of civic trust.

So, design is not neutral: it can clarify or obscure, empower or manipulate. This is where the book *Design for Privacy* comes in. It's an indispensable tool for any practitioner on the front lines who wants to ensure they are using their skills to build honest, respectful relationships between businesses, users, and their data.

—Harry Brignull
Expert witness and founder of Deceptive.Design

INTRODUCTION

Is your vacuum cleaner watching you? It could be. It could also be uploading photos, videos, and audios of you to train an AI model. "That's not what I signed up for, though!" you might say. Nonetheless, in late 2024, the Chinese robotics company Ecovacs was reported to be selling a model of its vacuum cleaner in Australia that did all of the above, while the company claimed that consumers were "willing participants in this AI training program."[1] However, users had to signal their consent via a mobile app for the vacuum cleaner that vaguely suggested users would be helping "strengthen the improvement of product functions" without explaining exactly *how*. The app mentioned a link to details about how this data collection worked. But there was no such link on the screen. Similar complaints have been made about iRobot vacuum cleaners that have uploaded inappropriate photos of their owners to the company's servers, which were later posted to Facebook.

Problems with privacy around people's data continue to grow, often in unexpected ways. For example, in 2023, Mozilla launched their first ever "Annual Consumer Creep-O-Meter," which concluded that the state of digital privacy then was "Very Creepy."[2] They also concluded that automotive was the worst category they had ever reviewed for privacy issues. Mozilla reported that Nissan collects "some of the creepiest categories of data we have ever seen," including your "sexual activity." Similarly, Kia mentions in their privacy policy that they may gather data about your "sex life." Six automotive companies reveal that they may be collecting your "genetic information" or "genetic characteristics." And in early 2023, former Tesla employees told Reuters that workers had shared videos taken by the

---

[1] Julian Fell, "Insecure Deebot Robot Vacuums Collect Photos and Audio to Train AI," ABC News, 4 October 2024, www.abc.net.au/news/2024-10-05/robot-vacuum-deebot-ecovacs-photos-ai/104416632

[2] "Mozilla's First-Ever Annual Consumer Creep-O-Meter," Mozilla, 2023, https://foundation.mozilla.org/en/privacynotincluded/articles/annual-creep-o-meter

vehicles internally, including "scenes of intimacy" and embarrassing clips, such as a naked man moving around his vehicle.[3]

These examples come from just two corners of the evolving knot of global privacy issues.

Developing an Eye for the Problem

We've all had that experience where, once you've seen something up close and in detail, you start seeing that thing everywhere. Maybe you bought a model of a car you'd never owned before—let's say a Subaru Crosstrek—and now, you'll see Crosstreks everywhere. That's what I hope to accomplish with this book, really. I hope to present you with certain patterns and examples of privacy issues, so that when you see them you jump and point. "I see that!" you might exclaim. "That's bad. Here's why. Here's how we could do better." If I can provoke that kind of reaction to privacy issues among designers, I'll consider this work a success. Then you may start to see privacy problems everywhere. Because they are, indeed, everywhere.

For many people, online privacy may never have proven to be a particularly troubling issue for them, personally. "I've got nothing to hide" is a pretty common reaction for many people, maybe even accompanied by a shrug. The National Association of Attorneys General describes this as "the granddaddy of anti-privacy arguments."[4] Disturbingly, they point out, the Nazi propaganda minister Joseph Goebbels even expressed the same sentiment, saying, "If you have nothing to hide, you have nothing to fear."

---

3   Steve Stecklow, Waylon Cunningham, and Hyunjoo Jin, "Tesla Workers Shared Sensitive Images Recorded by Customer Cars," Reuters, 6 April 2023, www.reuters.com/technology/tesla-workers-shared-sensitive-images-recorded-by-customer-cars-2023-04-06

4   Ryan Kriger, "Debunking the Privacy Fallacies," National Association of Attorneys General, 12 October 2021, www.naag.org/attorney-general-journal/debunking-the-privacy-fallacies

Edward Snowden, the former CIA employee turned infamous NSA leaker, may have become a polarizing figure, but he nails the issue this way: "Arguing that you don't care about the right to privacy because you have nothing to hide is no different than saying you don't care about free speech because you have nothing to say."[5]

So, even if you or I feel we have nothing to hide, as is the case with many rights, we don't always understand the need for privacy until it affects us, personally.

Even the most mundane, everyday use of your data has an impact upon your privacy by accretion. Globally, companies are tracking your digital exhaust every moment you're online, hoovering up your data to profit from it. Many of the experiences you use for free, of course, are collecting your data, repurposing it as their own, and selling it to third parties. Meaning, of course, that if something seems to be free on the internet, then it would appear that *you are the product*.

The tech writer Will Oremus disagrees with that conclusion.[6] In his 2018 Slate article, "Are You Really the Product?" he wrote:

> From a privacy standpoint, advertising and the internet were a match made in hell. The personalized-advertising model employed by Facebook, Google, and other online platforms is the product of that unholy union. And as a society, our acceptance of it amounts to something like a Faustian bargain. To the extent that our personal data has become a product, it's because we—and our representatives in government—have allowed it to happen.

Oremus concludes that if you agree that you are the product, that renders you helpless. Instead, he suggests you should be more assertive about fighting this received wisdom. I agree. And I think that we, as designers, have the opportunity, maybe even the mandate, to battle these dynamics on the front lines.

---

5 "Just Days Left to Kill Mass Surveillance under Section 215 of the Patriot Act. We are Edward Snowden and the ACLU's Jameel Jaffer. AUA," Reddit, 21 May 2015, www.reddit.com/r/IAmA/comments/36ru89/comment/crglgh2

6 Will Oremus, "Are You Really the Product?" Slate, 27 April 2018, https://slate.com/technology/2018/04/are-you-really-facebooks-product-the-history-of-a-dangerous-idea.html

So, if there's no mystery behind why corporations long for your personal information, you should still have the right to determine how much data you hand over to them and when. Business now treats your data as if it's their capital. I believe we can agree collectively—as consumers, as designers, as human beings, that we're not going to allow that.

Why Me?

What prompted my interest in this subject? On the one hand, it's difficult to pin down when I became interested in "privacy" in general and "privacy by design" specifically. On the other, there are a few strands that I recognize have come together to make this a topic that occupies a lot of my time and thinking.

There are many UX designers and consultants out here, just as there are many incredibly qualified privacy experts. I have simply found myself, often, at the axis of these two arenas and have spent a fair amount of time there now, dwelling on what I've seen.

Back in 2012, I wrote a *Medium* article entitled "Suddenly We're Ubiquitous" about how the preponderance of public photos and the increased use of and access to facial recognition would likely undermine our privacy, concluding that we had already "traded our privacy for dribs and drabs of entertaining data, for the ability to share ourselves." (People expressed fears about privacy when George Eastman debuted his then remarkable portable Kodak camera in the late 1880s, too.) At that time, I also noted a SXSW presentation by the Carnegie Mellon professor Alessandro Acquisti who demonstrated how with a single image you could likely determine someone's name and figure out most of their Social Security number. Needless to say, the sophistication of such technology and access to it have only increased in the intervening decade or so.

Some other elements of my personal background and experience certainly contributed to this interest of mine:

- Over a decade ago, I began working with different organizations that understood privacy to be of great value to their audiences.

I worked with and served on the board of BJUnity, a nonprofit that sought to assist LGTBQIA+ students at a notoriously strict private university. Due to my background, I understood the privacy issues faced by these students, who could get expelled from the school for their sexual orientation or gender identity, even if they didn't act upon it, and who studied and lived in an academic environment where their internet access—as adults—was monitored at all times, further compounding their anxiety.

- Similarly, I worked for several years with Day One, a nonprofit in New York City that assists young people who find themselves in abusive relationships and was enlightened as to their specific online privacy concerns, some of which pertained to their physical safety and the safety of Day One employees, too.

- Over the past decade, I've become increasingly involved in the internet freedom community, working on related projects and attending conferences globally. I've worked with designers, who focus on human rights projects to confirm that people who map to many different personas have intense privacy needs according to their unique situations: journalists, activists, protestors, members of the LGTBQIA+ community, women, and others.

- During the COVID pandemic I was a victim of theft and fraud via impersonation, and I can attest to the many hours of frustration and anxiety it caused me as I attempted to straighten everything out. This resulted from a banking security issue, certainly, but it also felt like a demoralizing assault on my privacy.

- I've been working in the field of user experience design for almost 25 years now, so I'm intimately familiar with how websites and applications use our data and monetize their ongoing existence. I'm more familiar than most with the design patterns they use—for better or worse—to acquire our data and what they proceed to do with it.

- Finally, like many who grow passionate about a particular topic, elements of my biography and upbringing certainly contributed to this interest, too. Those more personal details are probably best shared over a beer or coffee with those who are truly interested, but I have no doubt my lived experience contributed to my current focus upon privacy as a design issue.

If these experiences all helped me understand the needs of specific people and their individual threat models, I have also come to understand how privacy issues affect us all.

Who Is a Designer?

For the purposes of this book, I don't want to get too precious about who qualifies to be a designer. Every good or poorly executed, privacy-averse experience is the output of one or more designers, whether "designer" shows up in their job titles or not. So, although I am by occupation a user experience designer and consultant and the primary audience for this book may be UX or UI designers, I believe that, collectively, we gain nothing by limiting the conversation. If you're engaged in the creation of digital products, then this book is for you. You may be a UXer, an interface designer, a product designer, a product manager, a strategist, a developer, a producer, or a project manager. You're a designer.

This Book Is Biased

It's worth noting that many of the illustrations and examples contained in this book will likely represent a bias toward experiences designed and developed in the United States, such is the nature of my experience and the impact of this country, for better or worse, upon the world. I've tried to resist this bias, and, whenever possible, to cover examples of incidents and, especially, regulations that have more of a global impact. Nonetheless, I believe that the principles and best practices presented here will prove relevant globally. And if you have examples more relevant to your part of the world, I hope you'll forward them to me, so I can include them when I'm addressing this topic in the future. Similarly, it's been my hope to explain issues of privacy and solutions in ways that embrace more universal human values, not just the values of the environments I have found myself within.

This book is biased in another way, too. My background on this topic comes mainly from working within the user experience design community and, increasingly, within the internet freedom and human rights community. As such, this isn't a book that's written to ensure that designers stay merely within the boundaries of what is legal where they live. In fact, I'd argue that if designers merely adhere to the letter of the law, they could still be creating experiences that are actively harmful for people. Sometimes, experiences may be created to evade the letter of the law. In other cases, the law itself may be very forgiving of privacy issues or may even mandate experiences that undermine the dignity and privacy of users.

Consequently, although I've tried to avoid giving this book the tone of an activist's manual—and I may have failed at times—I'd happily admit it's intended to be a labor of advocacy. Not just an explanation of how to cover one's posterior from a legal perspective.

Ch-Ch-Changes

Finally, I want to acknowledge that for some months now I've been writing about a subject that is ever changing. Every day, I'd open a browser to find another piece of news related to privacy. A new law or court ruling. A huge fine levied against a well-known company. A new piece of technology that has the potential to trigger new consequences to privacy. Companies and their products appearing and disappearing or making sea changes to their business models.

I'm sure that the day after I hand over my final draft, I'll be alerted to a breaking news item or anecdote that I'll wish I could have included here. Everything is changing so quickly that some of what you read may not seem so fresh in a few years. And things won't stop changing. Still, even if we do find ourselves living in a bright and shiny future, where, say, cookie banners and privacy policies are relics of the past, I believe the principles you find here will still prove sound and helpful in designing experiences that help people to maintain their privacy.

CHAPTER 1

A Deluge of Privacy Issues

Issues with Data						2
Invasive AI						9
Cyberstalking and Bullying			12
What Can Designers Address?		13
The Takeaway						14

It's difficult to articulate our current circumstances without sounding like you're lapsing into hyperbole: We are currently experiencing an explosion of emerging privacy issues, unparalleled in human history that we can hardly expect the average person with a day job and a family and a mortgage to process, let alone keep up with. These issues are myriad, and they have a cumulative, if often unnoticed, effect upon people everywhere.

Before we get to the meat of this book, I want to offer a brief overview of some of these issues we're experiencing collectively to help highlight the scope of the problem—even if it means noting some issues that you, as a designer, may be incapable of addressing. Then we'll consider what you as a designer *can* potentially address.

Issues with Data

Much of the time, when you think of online privacy issues, you're thinking of data issues. You may find yourself asking, why are they asking for such personal information? Is my data going to be secure? Who is it being shared with? How can I prevent it from being shared if I don't want it shared? What should I expect if my information is leaked? And so on. You may find yourself cringing a little before joining a new experience and then proceeding with the shaky hope that you can trust the platform. Because there are a lot of problems that you can encounter with data.

Note, too, that some of these issues apply to the *security* of data but can have alarming subsequent impacts upon privacy, too. Everything is entangled.

Data Leaks and Hacks

Globally, users' data is increasingly exposed via a preponderance of data leaks. In April 2021, for example, Facebook, the largest, most popular social media platform on the planet was hacked. Subsequently, data for half a billion users—533 million to be specific—was leaked online, including people's personal information, such as their full names, phone numbers and email addresses, birthdates, and locations. That's precisely the sort of data that bad actors utilize to commit identity theft and fraud.

Early in 2024, a notorious hacker named USDoD announced that they had stolen the Social Security numbers of every United States citizen, after they hacked the data broker company National Public Data, stealing the records of some 2.9 *billion* people. A data breach of that scale could also ignite a firestorm of identity theft and fraud crimes.

In July 2024, too, AT&T announced that call records for almost all their customers had been stolen, potentially affecting over 100 million customers, but also anyone whom those customers called or texted. That leak included people's phone numbers and location data, but not the contents of any calls or messages. AT&T paid the hacker to delete the files. However, no one can be certain that data was completely deleted, and much could be inferred about individuals based upon their call records. As *The New York Times* pointed out:

> Careful analysis might reveal someone's political affiliations or sexual orientation based on the businesses and organizations they interact with. It could also show if someone has contacted abortion services or gender-affirming health care. That information could then be used for harassment, or possibly legal action depending on where the person lives.[1]

Again, these are security issues that designers can't necessarily solve for, but you can see how they could also have a profound effect upon people's privacy once specific elements of data are loosed into the wild.

Those are just three examples among thousands of data breaches that occur annually in the United States alone. Leaks like these are so common now that we almost shrug them off: Their scale may be too great to wrestle with.

---

1   Max Eddy, "The Massive AT&T Data Breach Doesn't Just Affect AT&T Customers. Here's How to Protect Yourself," *The New York Times*, 16 July 2024, www.nytimes.com/wirecutter/reviews/how-to-protect-yourself-att-breach

WHAT IS Q-DAY?

Imagine if all your passwords suddenly became useless to protect your personal information. In 2019, tech journalist Christopher Mims reported that security experts feared bad actors using quantum computers would break through the existing encryption technology that protects our data within a decade.[2] Those experts hope to pinpoint new ways to protect our personal information—and quickly. Cybersecurity researchers now fear generative AI will be directed to mimic biometrics, too, spoofing your fingerprints or your face, for example, and undermining what has been considered a powerful source of security for personal data.

In early 2025, Amit Katwala reported for *Wired* on this coming "Q-Day" in his ominously titled piece, "The Quantum Apocalypse Is Coming. Be Very Afraid."[3] "[On] Q-Day," he wrote, "everything could become vulnerable, for everyone: emails, text messages, anonymous posts, location histories, bitcoin wallets, police reports, hospital records, power stations, the entire global financial system."

If it came, Q-Day would be a massive security event, but one that would have profound implications for privacy in its aftermath. Contemplating it reinforces the point that companies should have good reasons to request certain personal information from users and that individuals should take care as to what they post online and pay close attention to their privacy settings for any experiences they're using. Note that designers can help in those two areas, too: Handling how data is requested and providing users with controls for how their data is used.

---

2 Christopher Mims, "The Day When Computers Can Break All Encryption Is Coming," Christopher Mims, *The Wall Street Journal*, 4 June 2019, www.wsj.com/articles/the-race-to-save-encryption-11559646737

3 Amit Katwala, "The Quantum Apocalypse Is Coming. Be Very Afraid," *Wired*, 24 March 2025, www.wired.com/story/q-day-apocalypse-quantum-computers-encryption

Data Sharing and Transparency

Perhaps the most common way that people's privacy is abused happens right under their noses every day: People are often not aware of the degree to which they're being tracked across the internet and to which their data and browsing behavior is shared with scores of other companies.

In the introduction, you learned how your vacuum cleaner may be watching you, and your car might be sending the data about your sex life to the automobile maker. But the demand for your data is far more pervasive, extending far beyond those two more outrageous examples.

The demand for personalized content fueled by a wealth of personal data seems higher than ever. People say they want personalized ads, so you'd think they'd enjoy sharing their data. If you let a company follow you around the internet, theoretically, you'll be served up better content and more targeted advertising. But things start to look very different when you ask people if they are OK with *how* personalization works, exactly.

In 2019, the network security company RSA found that only 17 percent of respondents believed it was ethical for companies to track their online activity to provide them with personalized advertising.[4] As far back as 2014, a Pew Research survey found that 91 percent of adults believed that "consumers have lost control over how personal information is collected and used by companies."[5] Additionally, 80 percent of respondents who used social networking sites said they were "concerned about third parties like advertisers or businesses accessing the data they share on these sites."[6]

In 2023, Publishers Clearing House Consumer Insights shared the results of their survey with 45,000 respondents.[7] It showed that 86 percent of Americans were more concerned about their privacy and data security than the state of the U.S. economy. Still two-thirds of Americans either don't grasp how their data is being used or how many entities have access to that information.

---

4   Ross Benes, "Do People Actually Want Personalized Ads?," 4 March 2019, www.emarketer.com/content/do-people-actually-want-personalized-ads

5   Mary Madden, "Privacy and Cybersecurity: Key Findings from Pew Research," Pew Research, 16 January 2015, www.pewresearch.org/short-reads/2015/01/16/privacy

6   Mary Madden, "Public Perceptions of Privacy and Security in the Post-Snowden Era," Pew Research Center, 12 November 2014, www.pewresearch.org/internet/2014/11/12/public-privacy-perceptions

7   Gary Drenik, "Data Privacy Tops Concerns for Americans—Who Is Responsible for Better Data Protections?," *Forbes*, 8 December 2023, www.forbes.com/sites/garydrenik/2023/12/08/data-privacy-tops-concerns-for-americans--who-is-responsible-for-better-data-protections

CASE STUDY

DATA AND ROE V. WADE

In early 2022, critics began voicing concerns that, if Roe v. Wade were overturned, personal data could be reviewed to pinpoint pregnant individuals, who might be considering an abortion.[8] This wasn't an entirely new concern. For example, in 2019 attorneys found that Missouri regulators had combed through publicly released but anonymized information about patients' menstrual periods from the last remaining Planned Parenthood clinic in the state hoping to uncover individuals who had undergone failed abortions.[9]

When Roe v. Wade was overturned, these concerns escalated. Within days Planned Parenthood announced they would remove Facebook's Meta Pixel from their site. This marketing tool allowed Facebook to track any visitors who visited the nonprofit's website. The online tech news organization, *The Markup* had been covering the use of the Pixel and other ad trackers since at least 2021. Days before Roe v. Wade was overturned, they released a study of 100 hospitals showing that 33 of them had inadvertently been sharing their patient's medical information with Facebook.[10] That meant that every time visitors to these sites scheduled an appointment, they unwittingly sent Facebook

---

8 Rina Torchinsky, "How Period Tracking Apps and Data Privacy Fit into a Post-Roe v. Wade Climate," NPR, 24 June 2024, www.npr.org/2022/05/10/1097482967/roe-v-wade-supreme-court-abortion-period-apps

9 Yasmeen Abutaleb and Emily Wax-Thibodeaux, "Missouri Reviewed Data About Planned Parenthood's Patients, Including Their Periods, to Identify Failed Abortions," *The Washington Post*, 30 October 2019, www.washingtonpost.com/health/missouri-tracked-planned-parenthood-patients-periods-in-spreadsheet-top-health-official-says/2019/10/30/e96791d0-fb42-11e9-ac8c-8eced29ca6ef_story.html

10 Todd Feathers et al., "Facebook Is Receiving Sensitive Medical Information from Hospital Websites," *The Markup*, 16 June 2022, https://themarkup.org/pixel-hunt/2022/06/16/facebook-is-receiving-sensitive-medical-information-from-hospital-websites

their data as well. Remarkably, *The Markup* also found Pixel planted *inside* several password-protected patient sites. *Vice* also reported on Placer.ai, a company whose data-enabled heat maps had allowed searchers to discover the approximate locations of people who had visited Planned Parenthood clinics.[11]

Immediately after SCOTUS overturned Roe v. Wade, privacy experts advised women and trans men to delete period tracking apps like Flo and Clue for fear they might eventually surrender sensitive data to law enforcement. The companies behind those apps responded by declaring their safety, promising they would continue to follow existing privacy laws. Flo soon announced they had begun work on an "anonymous mode" within their app to address these spiking privacy concerns.

But deleting period trackers may not be enough. Experts will remind us that emails, texts, internet searches, and website visits have all been used already to convict women on abortion-related charges. Authorities could use the location data from people's mobile devices to gather evidence against them, too.

This post-Roe fear offers just one reminder of the potential harm that can come to people when their data isn't handled respectfully or, alarmingly, when data offered by individuals for initially innocuous reasons, can later be used in ways that could prove profoundly harmful to them. It also reveals how privacy issues can emerge suddenly that affect already at-risk or disadvantaged groups. We should also remember that privacy issues can affect anyone, underscoring the need for us to consider specific best practices for both requesting and handling personal data carefully and responsibly.

---

11 Joseph Cox, "Location Data Firm Provides Heat Maps of Where Abortion Clinic Visitors Live," *Vice*, 5 May 2022, www.vice.com/en/article/location-data-firm-heat-maps-planned-parenthood-abortion-clinics-placer-ai

For example, how many third parties do you think some of the most popular websites people visit would share their data with? Dozens? Hundreds? *Wired* looked at the top 10,000 websites and found that thousands shared data with hundreds to well over 1,000 third parties.[12] One popular quiz site, JetPunk, shares data with over 1,800 "partners." Over 20 sites, including **Investopedia.com**, **People.com**, and **Allrecipes.com**, all owned by the publisher Dotdash Meredith, disclose that they may share data with 1,609 partners. Your eyes may glaze over when you see numbers like that as it's difficult to process the impact. In fact, Midas Nouwens, an associate professor at Aarhus University in Denmark who worked with *Wired* on the study, suggests that even an indication that data is being shared with more than five to ten partners becomes somewhat useless: "That's still too many for anybody to really form an opinion on considering how opaque and complex this whole data processing pipeline is."

If the mass-scale yet largely opaque sharing of data with business partners proves concerning, the practice of data sharing—intentionally or inadvertently—with law enforcement and other government bodies can prove positively alarming.

These data sharing issues impact people in different ways globally, too. For example, Egyptian police have been known to use dating apps like Grindr and WhosHere to entrap LGTBQIA+ people. Human Rights Watch reported dozens of instances where law enforcement in Egypt, Jordan, Lebanon, Iraq, and Tunisia harassed members of the LGTBQIA+ community after monitoring them on social media. Human Rights Watch criticized some social media companies for not providing better moderation and protection within their experiences.[13] And although Grindr warned people in Egypt that law enforcement may try to entrap them, Norway fined the company a record 65 million kroner ($7.16 million) at the time for sending users' personal information to hundreds of business partners without the user's consent.

Unfortunately, these examples represent the tip of the tip of the proverbial iceberg where data sharing issues are concerned. You could

---

12 Matt Burgess, "Some of the Most Popular Websites Share Your Data with over 1,500 Companies," *Wired*, 21 March 2024, https://wired.me/technology/the-most-popular-websites-share-data-with-over-1500-companies

13 Rasha Younes, "All This Terror Because of a Photo," Human Rights Watch, 21 February 2023, www.hrw.org/report/2023/02/21/all-terror-because-photo/digital-targeting-and-its-offline-consequences-lgbt

devote an entire book to this topic alone. Often, companies are not transparent about the data they use or who they're sharing it with. You can certainly advocate for improvement in this area as designers.

Invasive AI

The increasing use of technology, such as facial recognition and other forms of biometrics, tap into your personal data and enable innumerable companies, agencies, and authorities to monitor you in staggering scale and detail. Increasingly, if you live in a large city or metropolitan area, you may find yourself exposed to these technologies daily without even knowing it.

Facial Recognition

Clearview AI is a facial recognition platform that offers services to law enforcement and other government agencies. The company downloaded over three *billion* photos of people from social media sites, such as Facebook, Instagram, and LinkedIn, and used those images to build facial recognition models for millions of people globally without their permission. Clearview was hacked in 2020, and the stolen data revealed information about all their customers, including various law enforcement agencies, such as the police, the Department of Homeland Security, and the FBI. The company boasts they have over 50 billion images in their law enforcement database.[14]

Notably then, the use of facial recognition by law enforcement has led to several false arrests. In 2023, for example, police arrested an eight-months pregnant woman, Porcha Woodruff, and accused her of carjacking. They handcuffed her, took her to a detention center, where she spent 11 hours denying the charges, only for law enforcement to find they had made a false arrest due to a false match. At that time, Woodruff was one of six people to say they'd been falsely arrested because of facial recognition failures. All those individuals were Black.

Still, the U.S. government continues to divine new use cases for the technology. In 2024, *MIT Technology Review* reported that the Department of Homeland Security hoped to use facial recognition to identify migrant children, starting at infancy, in order to track them as they age.

---

14 Clearview site accessed 29 September 2024, www.clearview.ai

The private sector leverages this technology, too. Stores such as Ace Hardware, Albertsons, Macy's, and Rite Aid have used facial recognition programs to identify customers. Some use apps to track customers around their stores so they can present them with ads online later. Rite Aid was eventually banned from using facial recognition after the FTC determined they had launched the technology without putting the proper safeguards in place. The company relied on poor quality images, for example, that resulted, not only in false positives, but in false arrests, particularly of women and people of color.

The grocery store chain Kroger came under scrutiny in late 2024, too, for a plan to use both digital price tagging and facial recognition in their stores, which critics feared could lead to price gouging and charging customers varying prices according to their identities. Kroger and Walmart have been experimenting with digital pricing: Both companies claim they would not use it to enable price surging.

Misuse of Biometric Data

In early 2021, Amazon began requiring some 75,000 delivery drivers to sign consent forms that allowed the company to collect those driver's biometric data and to use AI cameras to monitor their location, movement, and driving patterns. If a driver so much as yawned, it triggered a camera to record their motions. Some drivers quit over this form of "AI surveillance." "It was both a privacy violation, and a breach of trust," one driver told Thomson Reuters at the time. "And I was not going to stand for it."[15] The manager for another driver told *Vice*, "It's a heart-breaking conversation when someone tells you that you're their favorite person they have ever worked for, but Amazon just micromanages them too much."[16]

---

[15] Avi Asher-Schapiro, "For This Amazon Van Driver, AI Surveillance Was the Final Straw," Thomson Reuters Foundation, 19 March 2021, https://news.trust.org/item/20210319120214-n93hk

[16] Lauren Kaori Gurley, "Amazon Delivery Drivers Forced to Sign 'Biometric Consent' Form or Lose Job," *Vice*, 23 March 2021, www.vice.com/en/article/amazon-delivery-drivers-forced-to-sign-biometric-consent-form-or-lose-job

DEEPFAKES AND VOICE SPOOFING

Various forms of generative AI now enable bad actors to spoof your face, your body, and your voice in ways that can prove destructive to your privacy and security, too.

When the war in Ukraine commenced, you may have seen deepfake videos debut featuring both Vladimir Putin and Volodymyr Zelensky. You've also likely seen them used for Presidents Joe Biden and Donald Trump, or even Pope Francis (see Figure 1.1), sometimes simply for humorous effect, but often for disseminating propaganda and misinformation, as well.

FIGURE 1.1 Many early viewers of this AI-generated image of Pope Francis seemingly decked out in a fashionable puffy coat believed it was real.

continues

A Deluge of Privacy Issues 11

> **DEEPFAKES AND VOICE SPOOFING** (continued)
>
> These deepfakes could, conceivably, trigger disastrous reactions. Increasingly, however, they can create havoc for everyday people, too. The distribution of deepfake nudes has become an increasing problem, and the technology is so readily available that teenagers have been arrested for creating deepfakes of their schoolmates. You may have heard any number of stories now of people receiving phone calls from a distraught relative, asking for ransom money because they've been kidnapped, only to discover they had been scammed by crooks using voice cloning to spoof their child or relative's voice. The FTC quickly outlawed robocalls using AI-generated voices, but stories like this may make you feel like you're living in a cyberpunk novel.
>
> We'll explore some extraordinary examples of how AI is undermining online privacy in Chapter 11, "The Evolving Impact of Privacy Policy," as well as guidelines for mitigating its impact.

Cyberstalking and Bullying

Implicit with issues such as Amazon's use of biometrics is the fact that people are being tracked via their mobile devices in ways unprecedented in human history. If you have a mobile phone and you use any apps or websites at all, your movements in the real world are constantly being tracked. So are the fine details of your online behavior.

As the details of our personal lives have become increasingly available online, cyberstalking and cyberbullying have increased, too. It's not unreasonable to conclude that these forms of harassment have risen along with the growth of social media. A September 2020 Pew Research Center study showed that 41 percent of Americans had experienced some form of online harassment, with about half of those saying they had been harassed due to their political beliefs.[17] Thirteen percent of women said they had been stalked online. Pew confirmed that these forms of online harassment had grown since 2014 and that more severe forms of harassment, including physical threats, stalking, sustained harassment, and sexual harassment had intensified since 2017.

---

17 Emily A. Vogels, "The State of Online Harassment," Pew Research Center, 13 January 2021, www.pewresearch.org/internet/2021/01/13/the-state-of-online-harassment

Research also seems to confirm a rise in cyberstalking and cyberbullying among university students and staff during the COVID-19 pandemic.[18] Calls to the National Stalking Helpline related to cyberstalking increased during the pandemic, too.[19]

When you are developing apps that encourage people to share their geographic data—even with family members—you must consider the potential for harm. For example, at Razorfish, our team had to consider what might happen if a couple using a car manufacturer's app broke up, but both still had access to each other's geographic location because they could see where a vehicle was at any time.

In Chapters 8, "Provide Tools for Enabling Privacy, and 9, "Cultivating a Culture of Privacy by Design," we'll look at two cases where well-known, heavily used apps might inadvertently become "stalkerware" because they debuted features that allowed people's geographic data to be made public by default.

> **NOTE SAFETY & PRIVACY**
>
> Just as considerations for privacy and security overlap, so do considerations for safety and privacy. Issues such as harassment, abuse, dogpiling, and online stalking that affect users' safety online feel like attacks on user's privacy, as well, especially when they lack tools to keep their accounts private or to block specific users from viewing their content. Safety by design is an approach or process itself with its own set of principles that you may want to investigate.

What Can Designers Address?

I mention all these issues to highlight the growing scale and diversity of the privacy issues we've been steeping in and, sadly, acclimatizing to. You can address some of these issues as a designer. Others, such as those more closely related to security breaches, you likely can't. You *can*, however, work to create experiences that help ensure users' security, and you can pledge to warn users about potential security issues involving their data.

---

18 Anna Bussu et al., "Exploring the Impact of Cyberbullying and Cyberstalking on Victims' Behavioural Changes in Higher Education During COVID-19: A Case Study," *International Journal of Law, Crime and Justice*, December 2023, www.sciencedirect.com/science/article/pii/S175606162300054X

19 "Unmasking Stalking: A Changing Landscape," Suzy Lamplugh Trust, April 2021, www.suzylamplugh.org/Handlers/Download.ashx

And, as this book will show, people consistently bring up an array of specific online privacy concerns that designers *can* address.

For example, they express frustration and even fear about the following:

- Companies that share opaque privacy policy changes they can't understand.
- Lack of transparency around how their personal data is collected and what happens with it afterward.
- Deceptive patterns that trick them into accidentally sharing their personal information.
- Experiences sharing their personal, sometimes intimate interests and preferences without their knowledge or consent.
- Apps and websites tracking their online activity and even their geographic location without their knowledge or consent.
- Experiences accessing, sharing, or reaching out to their contacts without their knowledge or consent.
- Experiences failing to offer them tools to control their personal data, delete or export their personal data, or cancel their accounts.

These are incredibly common privacy-related issues you *can* help prevent, or resolve, or, at least, mitigate in one way or another. They are real issues that affect real people.

The Takeaway

After studying these privacy-related experience issues for several years now, I've found that approaches for addressing these issues generally fall into a handful of categories.

Privacy issues can be addressed by considering one or more of the following key practices:

- Handle people's data responsibly.
- Avoid deceptive experience patterns.
- Advocate for better use of language.
- Provide tools that empower users to maintain their privacy.

Those approaches reflect four pillars for privacy by design. They form the core of this book. The pillars rest upon a foundation, too: a design culture better attuned to privacy concerns. Combined, these pillars and that foundation provide support for stronger, more authentic, and human-centered privacy user experiences.

CHAPTER 2

Defining Privacy

What's the Difference Between Privacy and Security?	18
Defining a Right to Privacy	18
Some Relevant Privacy Terminology	20
The Origins of Privacy by Design	23
The Takeaway	24

Privacy is a wriggly concept—difficult to pin down. In his seminal book on the broader topic of privacy, *Understanding Privacy*, Daniel J. Solove notes this difficulty in defining the concept.[1] "There is no overarching conception of privacy," he notes. "It must be mapped like a terrain, by painstakingly studying the landscape." He concludes that privacy "is a concept in disarray. Nobody can articulate what it means."

Still, if I'm going to predicate an entire book on the value of privacy by design, I should probably define it as clearly as possible.

Solove points out that privacy has typically been considered in one of six different ways. Privacy might be the right to be let alone. To limit access to oneself. To conceal information from others. To control your personal information. To protect your person. Or to protect your intimacy.

In the United States, that first definition—"the right to be let alone"—as framed by Samuel Warren and Louis Brandeis in their 1890 law review piece "The Right to Privacy," has probably had the greatest impact (see Figure 2.1). That definition proved so influential that it arguably set the stage for a right to privacy as well as privacy law within the United States for decades to come. The definition is so broad, however, that it doesn't really serve our purposes here.

FIGURE 2.1
These photos of Samuel Warren and Louis Brandeis bookend their influential essay "The Right to Privacy" for *Harvard Law Review*, 15 December 1890.

---

1 Daniel J. Solove, *Understanding Privacy* (Cambridge, MA: First Harvard University Press, 2009).

Daniel Solove concludes that no one definition of privacy fits all situations. Instead, privacy must be considered in different contexts and as differing across culture and time periods, too.

In the context of online activity, privacy arguably has a more market-driven approach with U.S. laws often evolving, so far, to enable businesses to utilize your data unless you ask them *not* to. The European approach treats privacy more like a fundamental human right. That latter approach aligns better, I believe, with the practice of privacy by design that we'll explore here. Also, consider that some societies work with a more individualistic emphasis upon addressing privacy problems, while others emphasize a more collective approach, treating privacy as a societal issue to be addressed, like poverty or climate change.

For the purposes of this book, my definition is specific to the context of online privacy and user experience design. It's closest to that fourth type Solove mentioned previously. I'd like to define privacy as simply as possible: Privacy is your right to maintain control over your personal information and preferences.

Now, that element of "personal information" covers a lot of ground potentially, and, depending on the context, could include information such as your name, age, weight, gender, likeness, address, phone number, email address, geographical location, Social Security number, marital status, medical history, sexual orientation, political alignment, income, and on and on. It could include your correspondence and communications, as well. And that second element of "preferences" could envelop just about anything you indicate that you like or prefer, since you cannot always know the quite legitimate reasons someone might have for wanting to keep a specific preference to themselves. For one person, having their interest in a particular advocacy group like the Trevor Project trumpeted to the public may not be a big deal. For another, that sort of exposure could be disastrous. The various complex combinations of these data points and preferences and the resulting privacy concerns they present are as myriad as the people they represent. Additionally, people may feel perfectly fine sharing their personal information in one context and not in another. And that's perfectly fine, too!

So, I've arrived at a simple definition. But it's complicated.

DEFINING PRIVACY

What's the Difference Between Privacy and Security?

It's easy to talk about privacy and security as if they are interchangeable concepts. They are different, if related, concepts, so let's pinpoint the difference between these terms before proceeding any further.

For our purposes, let's consider the two concepts this way:

- *Privacy* speaks to *your* ability to control your own personal information and how it's accessed or used.
- *Security* considers how other entities protect your personal information—those companies or organizations that hold onto your data, and, quite importantly, may share it with other parties or even the public (intentionally or not). Of course, you play a role in protecting your own information, too.

These concepts often do overlap, so you'll frequently find both referenced, although our focus will be on considering how as a designer, you can ensure people's privacy is maintained as you create experiences for them.

> **NOTE** **THE ROLE OF INDIVIDUALS IN SECURITY**
>
> Individuals do play a role in maintaining the security of their data, too, whether by updating their passwords, storing their passwords securely, or using two-factor authentication. For more on that topic, see Heidi Trost's book, *Human-Centered Security*, which looks at the security ecosystem itself to consider the roles of users, threat actors, and the design system itself.

Defining a Right to Privacy

We human beings share the extraordinary ability to imagine, claim, and then develop rights for ourselves. Among these, privacy is well established as a universal human right.

In 1928, U.S. Supreme Court Justice, Louis Brandeis declared privacy to be "the right to be left alone—the most comprehensive of rights and the right most valued by civilized men." Although the word "privacy" does not appear in the Fourth Amendment to the U.S. Constitution, the Supreme Court has repeatedly ruled that it provides for a "reasonable expectation of privacy." The Supreme Court

Justice John Marshall Harlan II coined that expression in a ruling on Katz v. United in 1967, creating the Reasonable Expectation of Privacy Test that is now considered to be the primary development resulting from that case.

In an article on the National Association of Attorneys General website, Ryan Kriger, Assistant Attorney General for Vermont, also says that the Supreme Court has confirmed that the right to assembly includes a right to privacy, as do conversations between spouses or congregants and their ministers.[2]

Kriger's conclusions around privacy fall firmly in favor of the individual:

> A *right* doesn't have to be justified—it is accepted as a quality of being a citizen or a human being. Entities that want to spy on you, read your communications, film your comings and goings, track your movements, and compile dossiers on you are defying accepted norms. Those entities are the ones that should be justifying *their* antisocial behavior, not the individual seeking to protect their privacy rights.

So here in the United States, privacy proves to be more than just a concept: It's explicitly a right.

Further, according to the International Association of Privacy Professionals, some 137 countries globally had declared privacy as a right within their borders by early 2024.[3]

Article 12 of the Universal Declaration of Human Rights focuses on a universal right to privacy and was adopted by the U.N. General Assembly in 1948.[4] It states:

> No one shall be subjected to arbitrary interference with his privacy, family, home or correspondence, nor to attacks upon his honour and reputation. Everyone has the right to the protection of the law against such interference or attacks.

---

2 Ryan Kriger, "Debunking the Privacy Fallacies," National Association of Attorneys General, 12 October 2021, www.naag.org/attorney-general-journal/debunking-the-privacy-fallacies

3 Aly Apacible-Bernardo, "Identifying Global Privacy Laws, Relevant DPAs," International Association of Privacy Professionals, 19 March 2024, https://iapp.org/news/a/identifying-global-privacy-laws-relevant-dpas

4 "Universal Declaration of Human Rights," United Nations, www.un.org/en/about-us/universal-declaration-of-human-rights

The global understanding of this right to privacy gives us some pretty significant motivation—and ammunition if need be—when considering how and why to design experiences with privacy in mind.

Some Relevant Privacy Terminology

While you're focused on terminology, let's not beleaguer you with too much privacy and security jargon because this is intended to be a design book, not a privacy and security manual (though, hopefully, this book will appeal to people working in that arena, as well). Still, you should acquaint yourself with some of the key terminology, relevant to the topics of privacy, security, and data processing in general.

First, consider *personally identifiable information* (PII)—PII consists of some of the most important information that defines you as an individual human being. It's generally considered any single element of data that can pinpoint you, specifically, such as your Social Security number, phone number, email address, driver's license number, passport number, patient ID number, or financial account number, but could also include your full name (which may not be completely unique) or other information. These data points are incredibly important because, when combined, two or three of them could be used to steal your identity or to confirm your identity for other reasons. Although other key data points, such as gender, ZIP code, place of birth, religion, or date of birth are not typically considered PII, they can certainly assist bad actors (criminals, oppressive government authorities, etc.) in quickly and effectively pinpointing you when combined with elements of PII.

Also, what is considered to be PII can change over time, as technology evolves, so that data points such as an email address or even a device ID or IP address would now be considered PII in some contexts, as well (in California and under the EU's General Data Protection Regulation (GDPR), for example). Some definitions of metadata might even include metadata or their browsing behavior, too, especially when they can be linked to an individual.

Additionally, when considering the privacy and security of people's—or companies'—data, you'll encounter the following terms that reference who owns or who is handling that data:

- **Consumer:** The consumer is the person who owns their own data and maintains a right to restrict the use of it, especially in the context of purchasing products and services. Of course, not everyone who wants to maintain control over their personal data meets that definition of "consumer," but the laws of some regions combine both types of people. (See California's Privacy Rights Act, for example.)
- **Controller:** The controller determines how and why a consumer's personal data is being handled, including how it's collected, used, and stored. They also determine the legality of handling that data.
- **Processor:** This entity processes data on behalf of the controller. They have to abide by any restrictions provided by the controller, too. A processor could take many forms, such as a *customer relationship management* (CRM) software company, a cloud storage provider, a payroll provider, a marketing company, or any other company offering services to the controller.
- **Data broker:** A broker collects data from various sources and creates profiles based on this data, often to sell to third parties—possibly without the consumer's knowledge.

For example, if you buy a pair of shoes online from Nike, that company may sign you up for a newsletter that is created on a CRM platform. In this scenario, you would be the consumer, Nike would be the controller, and the company owning the CRM platform would be the processor. If Nike also sells your data to a third party to create personalized advertising elsewhere on the web, then that company would be the data broker. And many, many companies could be leveraging the data from that broker to offer up ads to you (see Figure 2.2).

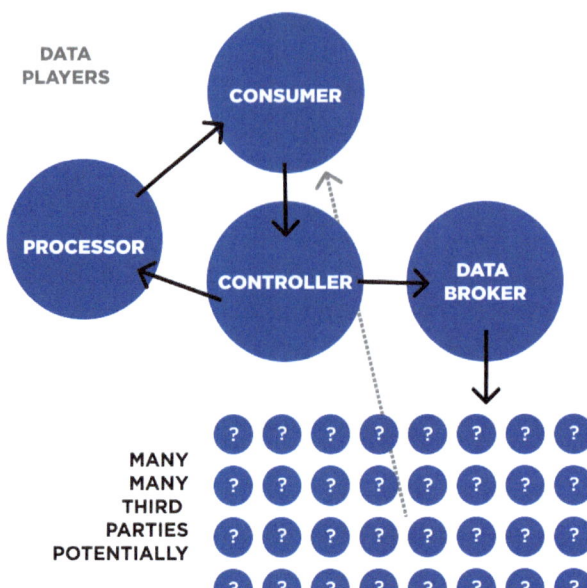

FIGURE 2.2
This diagram depicts how the four primary parties interact with an individual's data, as well as the many third parties, who data may be shared with without the consumer's knowledge.

Additionally, some huge companies like Amazon, Apple, Facebook, and Google may operate as controller, processor and data broker, all at the same time. Further, a company like Google may not need to sell user data to third parties. Instead, they can simply display personalized advertising for Nike, based upon what Google already knows about you.

> **NOTE** **THE PRIMARY PARTIES INVOLVED WITH PRIVACY**
>
> Just as those parties are involved in generating, sharing, and selling data, from another perspective, you can consider the parties most prominently involved in the broader world of privacy by design. Those primary players would be designers, users, business and policymakers, and regulators. These parties interact or conflict in different ways at different points in time, and the dynamics created by their interactions determine how everyone accesses experiences and how those experiences are updated over time to reflect the actions or demands of all parties involved.

The Origins of Privacy by Design

The ideas behind privacy by design can be traced back as far (at least) as the 1970s. However, in 1995, the work of the following three parties was formalized in a report: Dr. Ann Cavoukian of the office of Information and Privacy Commissioner of Ontario, the Dutch Data Protection Authority, and the Netherlands Organisation for Applied Scientific Research. Arguing for privacy by design as a practice, they essentially called for privacy to be a concern throughout any systems engineering process, rather than leaving it as a last-minute consideration after such a system has already been developed. The best practices or framework for this privacy-enhancing form of design were codified in what's now known as *Dr. Cavoukian's Seven Foundational Principles for Privacy by Design*.

Rather than revisiting each of those principles in detail in this book, I will spend much more time on the specifics of "how" to design for privacy in ways that support those principles. Our focus will be upon learning to recognize designs that fail to address privacy concerns by reviewing real-world examples and examining those best practices best suited to remedy any issues.

> ### THE SEVEN FOUNDATIONAL PRINCIPLES
>
> The Seven Foundational Principles for Privacy by Design from Dr. Cavoukian's framework follow:[5]
>
> 1. **Proactive Not Reactive; Preventative Not Remedial**
>
>   Anticipate, identify, and prevent privacy invasive events before they occur.
>
> 2. **Privacy as the Default Setting**
>
>   Build in the maximum degree of privacy into the default settings for any system or business practice. Doing so will keep a user's privacy intact, even if they choose to do nothing.
>
>   *continues*

---

[5] "Privacy by Design," Information and Privacy Commissioner of Ontario, January 2018, www.ipc.on.ca/sites/default/files/legacy/2018/01/pbd-1.pdf

THE SEVEN FOUNDATIONAL PRINCIPLES (continued)

3. **Privacy Embedded into Design**

   Embed privacy settings into the design and architecture of information technology systems and business practices instead of implementing them after the fact as an add-on.

4. **Full Functionality—Positive-Sum, Not Zero-Sum**

   Accommodate all legitimate interests and objectives in a positive-sum manner to create a balance between privacy and security because it is possible to have both.

5. **End-to-End Security—Full Life Cycle Protection**

   Embed strong security measures to the complete life cycle of data to ensure secure management of the information from beginning to end.

6. **Visibility and Transparency—Keep It Open**

   Assure stakeholders that privacy standards are open, transparent, and subject to independent verification.

7. **Respect for User Privacy—Keep It User-Centric**

   Protect the interests of users by offering strong privacy defaults, appropriate notice, and empowering user-friendly options.

The Takeaway

When you're designing for any complex experience, you might find it daunting to keep all these players and principles in mind, but, rest assured, exposure to these dynamics will help you recognize them much more readily in the days to come.

Now that you understand the scope of privacy issues, as well as our contextual definition of privacy and the players involved in privacy as an issue, let's get down to brass tacks about how you can help people avoid these issues in the first place.

CHAPTER 3

Your Role as a Designer

Privacy and the First Ethical Hackers	26
History May Not Repeat but It Rhymes	29
Why Me?	31
Who Are These "Users" Anyway?	33
The Takeaway	35

Good design can mean the difference between life and death. Consider the automobile. The first cars didn't have seatbelts or airbags. They certainly didn't have four-wheel anti-lock braking systems. Still, as each of those thoughtfully designed and increasingly sophisticated inventions has debuted, they have saved lives. And, eventually, automobile manufacturers were required to include each of these safety devices by law.

What about digital design, though? Can it save lives? Can it harm lives? As you'll see, digital design can do both. Experiences can incorporate design patterns that intentionally or inadvertently harm people. But they can also incorporate thoughtful patterns to help and even protect people, too. Fortunately, you, as a designer, are well positioned to help ensure the latter. In fact, arguably, you're better positioned than most people.

Privacy and the First Ethical Hackers

In the early 1940s, a French civil servant, René Carmille volunteered to spearhead the French census (see Figure 3.1). He and his team would provide detailed personal data about French residents to the Nazis. Carmille is known for pioneering the use of punch cards with early IBM systems for civil registration and for creating what would become France's 13-digit Social Security number, which he based on key data points, such as an individual's location, trade, and their specific skills.[1] Somewhat paradoxically, he and his team have also since been dubbed the world's first "ethical hackers." The reality of how they became hackers in the first place, however, proves much more complicated.

FIGURE 3.1
René Carmille deceived the Nazis while leading the French Census, and protected countless lives during World War II.

---

1   Edwin Black, *IBM and the Holocaust* (Crown Publishers: New York, 2001).

Carmille initially went to some lengths to assist the Germans by offering to provide census information that would help them pinpoint the identities of Jewish people living in France. They trusted him to the degree that they waited, repeatedly, for his results, not grasping that he was intentionally dragging his feet and inventing obstacles to avoid surrendering the data they coveted. They did not suspect he was also a member of the French Resistance.

In fact, given the horrifying, escalating activities of the Germans in Europe, Carmille and his team had moved their tabulating machines to a secret garage. Then he and his team decided to sabotage their own census work by ensuring that the punch cards never registered a result for column 11, which recorded someone's nationality (see Figure 3.2). They ensured that the processing and tabulation of over 100,000 cards highlighting Jewish individuals was never completed.

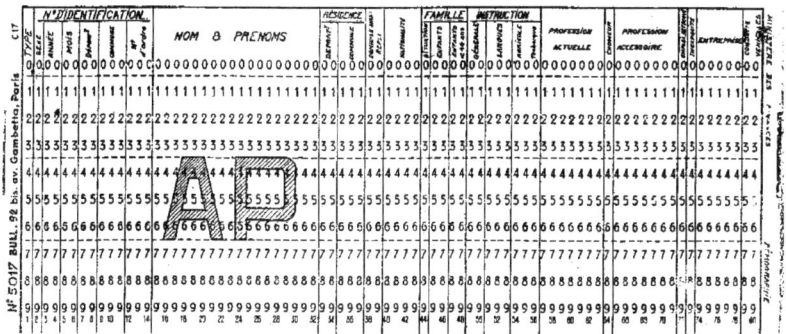

FIGURE 3.2
The "AP" paper punch card used in Vichy France to track French residents for the census. The "AP" stood for "Activités Professionnelles" or "Professional Activities."

Eventually, however, the team was discovered, apprehended by the Nazis, and tortured. For two days, Carmille himself suffered at the hands of the notorious SS officer Klaus Barbie (aka "the Butcher of Lyon"). He gave up nothing. Consequently, he was deported from his home country and imprisoned at Dachau, where he died of typhus in January 1945, just four months before the Germans surrendered to the Allies in Reims, France.

Although their efforts were suspended, Carmille and his team prevented the Nazis from discovering the identities of tens of thousands of Jewish people living in France, undoubtedly saving their lives.

There's a tragic irony to this story, however, when you consider that Carmille led the creation of the national identification number, the very intention of which was to track French residents. In other words, he first created something that already arguably undermined people's privacy. He eventually sabotaged the system when he saw how it could be misused. He even made his conclusion explicit, when he wrote of the National Statistical Service (SNS) that "its purpose is purely statistical, excluding any police role."[2] Others, of course, applied a different set of values to the system. The resulting experience would be used to hunt and terrorize people.

That should resonate with us as designers in two ways: First, take great care in designing experiences to avoid creating experiences that could be used to endanger people—even if that harm is the result of unintended consequences. Second, Carmille and his team saved those lives because they were committed to disrupting the very technology they designed to collect personal data, so they could preserve some particularly vulnerable people's privacy.

The point here isn't to advocate for sabotage—except in the direst of situations. Let's hope that's not a solution you'll ever need to consider. Your mission then may be less dramatic but still significant: Help organizations understand the implications of mishandling data and work with them to fix problems when there are lapses.

---

[2] Pierre Piazza, "The Identity Registration System, Identification Number and National ID Card During the Vichy Regime (France, 1940–1944)," *Criminocorpus*, 2017, https://journals.openedition.org/criminocorpus/3659

OTHER WAYS OF PUSHING BACK

Companies sometimes find constructive ways to make it more difficult for authorities to access data that could be used to harm individuals—which avoids the need for sabotage. For example, Apple introduced an "inactivity reboot" feature that causes iPhones to reboot if they haven't been unlocked for four days. The device then enters a more secure "Before First Unlock" mode, requiring a PIN, making it more difficult for thieves—and the police—to access someone's personal data. Android devices share this feature now, too. Apple has also denied government requests to build backdoors into their devices that would allow law enforcement to have access to people's personal data.

After 9/11 and in response to the Patriot Act, some libraries began informing patrons that their reading habits might be scrutinized by the FBI. (As far back as 1972, a librarian Zoia Horn was jailed for almost three weeks after she refused to share information about anti–Vietnam War activists.) Similarly, companies like Apple have engaged in the subtle practice of "warrant canaries." If the government puts them under a gag order, they conclude they can't trumpet the fact that the authorities have requested consumer data from them. Instead, they may publish a statement periodically to indicate that they have *not* received any demands for data in the past, say, six months. If some time elapses, and a similar statement fails to appear, you might rightfully infer that the government has since ordered the company to surrender such data.

History May Not Repeat but It Rhymes

Recent history has shown that employees working with user data and designing for systems and user experiences have also determined that they needed to take a stand against some remarkable attacks on people's privacy.

For example, in 2019, five employees quit their surely lucrative jobs at GitHub, the San Francisco tech company, after learning that the company shared data with Immigration and Customs Enforcement or ICE, the government agency that has been accused repeatedly of human rights violations—especially with regard to the treatment of undocumented immigrants, including enforcing family separations and placing immigrant children in cages. Some 150 employees even contributed to a letter to their company, insisting that Github end its $200,000 contract with ICE. In response, GitHub's CEO pledged to donate all the proceeds of the contract, as well as an additional $500,000, to immigrant-focused nonprofits. Similarly, in 2023, a former Twitter employee, Steve Krenzel revealed what he called "the most unethical thing I was asked to build while working at Twitter."[3] (See Figure 3.3.) He said the company had been approached by a large telecommunications company that, after some cautious hedging, said they wanted to know how many of Twitter's users were entering their competitors' stores.

Steve Krenzel
@stevekrenzel

With Twitter's change in ownership last week, I'm probably in the clear to talk about the most unethical thing I was asked to build while working at Twitter.

2:26 PM · Nov 7, 2022

FIGURE 3.3
The first tweet in Krenzel's 2022 thread about his experience pushing back on a telco's request for detailed tracking data for Twitter's users.

Krenzel tried to find a safe, anonymous way to divulge this information, while still respecting users' privacy, but his solution didn't satisfy the company. Instead, a frustrated director from the telco finally cut to the chase, saying, "We should know when users leave their house, their commute to work, and everywhere they go throughout the day. Anything less is useless. We get a lot more than that from other tech companies." Krenzel refused. His new manager eventually came to him, asking, "If we filled a dump truck with money and dumped it on you, would you stay and build this?" His response? "No dice."

---

3   Steve Krenzel, Twitter, 7 November 2022, https://twitter.com/stevekrenzel/status/1589700721121058817

Krenzel also relayed his concern to Jack Dorsey, Twitter's CEO at the time, who agreed that the social media platform should not cooperate, saying, "Let me look into that and make sure there isn't a misunderstanding. It doesn't seem right. We wouldn't want to do that."

Here Krenzel played an instrumental role in preventing another company from misusing Twitter users' personal location data in ways most of those individuals likely never even contemplated.

Why Me?

You may never be faced with dire choices like these as a designer, but you may find yourself facing down any number of micro decisions that will affect people's privacy, and these moments can add up, leading to more significant and unexpected privacy violations. So, even if you never have to take a dramatic stand that makes international news, you may well find yourself having to address issues on a smaller scale that still do affect people's privacy.

What is your role in all this as a designer likely to be then? Realistically? Practically? Why you?

Your Unique Position

As a designer, you are uniquely positioned to speak on behalf of users when their privacy is affected. After all, it's your job to keep users in mind at all times, right? But, more specifically, too, it's your role, not only to consider every part of an experience in comprehensive detail, but also to review dozens, even hundreds of minuscule but impactful design decisions with your company or clients. Often, you're interacting with them in person, too, or at least, via a Zoom or Teams call. You act as the hinge between the requirements of the stakeholders and the needs of the user. And you're often given the most airtime to articulate why each element of an experience should behave the way it does. Or to explain why it should not. You are the subject matter expert.

In his book *Ruined by Design*, Mike Monteiro advocates for designers to feel empowered to act on behalf of users. "You were not hired to get approval or to have your work pinned to the company fridge," he says. "People hire you to be the expert, so you might as well be the expert."[4]

Your experience certainly may vary according to the environment you're working within: Some companies have more developed design cultures. You may find that within other companies it's difficult to have much of an impact at all or even to provide feedback to the right audience. Also, admittedly, you may need to muster some confidence to address these issues head on, and confidence often comes with experience. Obviously, too, critiquing your client or organization's requirements requires a measure of tact and diplomacy. Those skills come with experience, too. If you're a more junior designer, you may feel neither confident nor empowered by your team's leadership to address these issues explicitly with your stakeholders. You may be afraid of putting your proverbial foot in your mouth. That's OK. Table your concerns in the moment, if need be, and then consider how you can address your concerns internally with your team later.

Your Particular Responsibility

As a designer, you have a particular responsibility. Even if, as an individual, you're not concerned with a particular privacy issue, personally, remember you're not designing for yourself. In his book, *The Descent of Man*, the British artist Grayson Perry reminds us that design has long been a job populated mostly by men, many of whom suffer from what he calls "self-as-user syndrome."[5] "That is," he explains, "in a typical masculine way they lacked empathy for an average user, especially if she was female." Of course, we're all capable of harboring biases as human beings, so I'd suggest we adopt Perry's term for this affliction into our design lexicon immediately: "self-as-user syndrome."

The word *empathy* comes up often in discussions about design. You might reference *empathy* because you are aiming to take your thinking a step beyond that incredibly low hurdle of "what works." You aim,

---

4   Mike Monteiro, *Ruined by Design* (Mule Design, 2019).
5   Grayson Perry, *The Descent of Man* (Penguin Books, 2016).

I hope, to do better than just "move fast and break things." You aim, insofar as possible, to put yourself in the shoes of other people, so that the experiences you design prove safe and comfortable for them. If you're designing with empathy, you'll consider the needs of people who are not like yourself—people with different backgrounds and experiences. That means not only researching privacy issues from a detached distance but also immersing yourself in the field and interviewing or talking to people with diverse backgrounds and lived experiences. You can't pretend to even approach that trait of empathy without some measure of thoughtful consideration and research. You need to get to know your users—human beings whose needs vary and need to be considered in varying amounts appropriate to very specific experiences. This practice gets to the very heart of user- or, better, human-centered design.

Some in the design community have argued that you shouldn't pretend to attain empathy because you can't possibly put yourself in another person's shoes. That's a fair point. But you can aim, at least, to be more understanding of people's diverse lives, needs, and situations, so that you design experiences that best meet their needs. And, if you try to cultivate this understanding for your users, you will be much more likely to address privacy issues that may affect them.

Who Are These "Users" Anyway?

You should make this distinction, too: Not all user experience design is *user-centered* design. You could even imagine some form of "user-centered" design that is studiously researched but focused on creating awful but increasingly efficient experiences for human beings. It's entirely possible to create experiences that intentionally deceive and ensnare users, and, I suppose, you could write an entire book that serves to divulge and advocate for those practices. This is not that book. However, we will examine some of those practices to consider why you should resist them and how you can design intuitive, usable, and transparent experiences that strive to please users and to maintain their privacy.

For the purposes of this book, let's assume that the practice of user-centered design leads to the design and development of inherently better experiences. Let's assume you're practicing truly *human-centered* design.

It's worth taking a moment to remind yourself who these "users" *are*, and who you have agreed you have a responsibility to advocate for. That term *user* can seem terribly abstract. Useful as it is, the word tends to strip people of their individual circumstances, their personality, their idiosyncrasies, and, perhaps most importantly, their lived experiences. It becomes a little too clinical, putting us at a stark remove from the lives of others.

Referencing the sort of data tabulation the Germans used to pinpoint individuals during World War II, one British intelligence report concluded, "The human being [now] becomes a number."[6] The stakes don't have to be that high for you to understand: You cannot afford to reduce your users—your fellow human beings—to numbers. That mountain of user data you may find yourself combing through for a project represents a totality of individual human lives.

As these individuals live their lives, too, new scenarios can arise that create new user needs. And with these new needs come new personas or archetypes and new threats for those users, too. In the wake of that spike in identity theft during the COVID-19 pandemic, for example, privacy and security issues became much more relevant to many more people. As we've seen, seismic cultural changes, such as the overturning of Roe v. Wade, have made privacy more relevant to people in frightening new ways, too. As a result of unfolding events, both here in the United States and globally, vulnerable groups such as women, transgender people, and undocumented immigrants have had to adjust their threat models in anticipation of the harm that now comes from their interacting with otherwise innocent-seeming digital products.

Keep this in mind then: You have a responsibility to real, flesh-and-blood human beings. *That's* who you're designing for. Given that responsibility, you may sometimes find that you need to push back on directives that adversely affect those human beings in terms that business understands. More on that in the next chapter.

---

6   Black, *IBM and the Holocaust*, p. 331.

The Takeaway

It's worth considering what you would do if faced with a situation where you determined that harm may be inflicted upon users' privacy if you proceed as you've been instructed to. The stakes won't likely be as high as they were for René Carmille and his team. But how would you proceed if you were asked, say, to create a deceptive user pattern that tricks users into surrendering their personal information, instead of asking for it openly? (That's a *much* more likely scenario.) What if you were working on a project and found that the app or website you're designing shares sensitive data or browsing behavior with particular third parties that may have an interest in harming people, and that fact has not been highlighted for users?

You might find it anxiety-inducing or even frightening to speak up in such situations, but (hopefully) you got into this business to help people—and you understand that the experiences you design for them will have a real-world impact.

Let's reinforce this point one more time, too: That "user" on the other end of the internet is a living, breathing human being. *That's* who you're designing for. Finally, keep this in mind: You are not alone. Collectively, as designers, we are well positioned both to align on this issue and to take the lead in emphasizing the importance of privacy by design.

CHAPTER 4

Why Should Business Care?

Civic Responsibility	38
Damage to Reputation	39
Experience Abandonment	40
Loss of User Base	41
Penalties and Fines	44
The Takeaway	46

After presenting on this topic, I'm sometimes asked a fair question something along these lines: Unless there are some sort of regulations in place, how can we possibly expect companies to care about these privacy issues—especially when, from their perspective, the tactics they're using may often lead to successful outcomes? As you'll see, companies certainly do need to be aware of the emerging regulations that may force their hands on these issues, but there are other reasons for them to consider consumers' privacy concerns.

Brands—your clients, the company you work for—should consider the privacy of their users' data, their content, and their browsing behavior for their benefit and safety. But let's be real: If they do express any concern at all, they often do so for their own institutional and financial self-interest.

Given that reality, you may be compelled to remind your clients of the potential impacts of ignoring their users' privacy concerns. So, what factors, specifically, might motivate businesses to pay attention to these concerns? Let's review five reasons for businesses and organizations to cultivate and maintain respect for people's privacy.

Civic Responsibility

Let's start with the most high-minded rationale for attending to your users' privacy concerns: the one grounded in ethics. Unfortunately, this rationale probably also happens to be the easiest for many organizations to ignore. As a user-centered designer, you should encourage your clients to treat their "end users" as real, flesh-and-blood human beings—members of their real-world community. That means people like their own friends and family, their coworkers and colleagues. How you treat those users impacts them in ways that have concrete moral and ethical implications.

Think of this as the argument for simply "taking the high road." To borrow from Michelle Obama's admirable (if challenging) maxim, you might advise business to take the approach of "When our competitors go low, we go high." Yes, that's an appeal to the heart and conscience, not the wallet, but you can advocate for empathy as an ideal characteristic for interacting with your users.

If this rationale sounds a little too idealistic for some, there's also a pragmatic or company-centric reason to adopt this approach: Demonstrating these characteristics of empathy and integrity simply contributes to a more palatable brand identity. Consider how Apple has enhanced their brand by placing an emphasis upon privacy and releasing features like App Tracking Transparency that help users maintain their privacy. The better the brand reputation, the better the opportunities for business. You might wish that business would act out of the goodness of its heart, but, at least, if it takes this more cynical approach, the effects are the same.

Damage to Reputation

That speaks to reputation, too. If a business is less interested in handling privacy matters with a purely ethical motivation, it is more likely to consider those developments that can damage their reputation. You might need to remind your clients that the way companies choose to handle privacy matters can undermine their brands.

Reputational damage suffered from privacy issues can seem nebulous and be difficult to measure, and larger companies may figure they can get away with something in the short-term and that their brand will prove resilient in the long-term. Still, a 2012 IBM study concluded that brand and reputational fallout from data breaches can last anywhere from months (enough to affect those all-important quarterly numbers) to years.[1]

Brand experts agree that when Elon Musk changed the name of the microblogging platform from Twitter to X, that change took a sledgehammer to the brand, measurably sending its value spiraling downward to the degree that the company has, for now, fallen out of global brand rankings. The brand valuation and strategy consultancy Brand Finance concluded that X is suffering a "major reputational crisis that the brand is struggling to overcome."[2] Further, a series of seemingly erratic changes to the platform's user experience have undermined both safety and privacy for users of the platform, which in turn has contributed significantly to that brand collapse.

Specific impacts on a brand's reputation can also include an erosion of trust and negative word of mouth. You have, no doubt, seen

---

1 "The Economics of IT Risk and Reputation," IBM, July 2013 , pdf, https://silo.tips/download/the-economics-of-it-risk-and-reputation

2 "The Decline of X: Musk's Rebrand Wipes Billions in Brand Value," Brand Finance, 12 September 2024, https://brandfinance.com/press-releases/the-decline-of-x-musks-rebrand-wipes-billions-in-brand-value

discussion around these dynamics before: People share their negative encounters with brands more often than their positive ones. And they tend to share those stories with more people, as well. Former Googler and privacy expert Tal Herman shared a Dutch saying on this point: "Trust comes on foot but leaves on horseback."

Experience Abandonment

Requiring users to share unexpected personal information, or using deceptive patterns intended to trick them into sharing personal data, may anger people and cause them to abandon your site in the heat of the moment in favor of another with a more transparent, privacy-friendly experience. Companies should ask themselves then: Is deploying that deceptive pattern or requiring that unnecessary personal data point really worth the increased exit rate?

For example, a 2018 study found that 56 percent of customers will abandon their carts when they discover shopping costs they weren't expecting.[3] Content marketer Rudy Klobas has described deceptive patterns as "the ultimate conversion blocker for ecommerce sites."[4] These patterns erode trust, he explains, "They're manipulative and seek to take advantage of those using your website. Using them puts you in the category of the shady used car salesman who does everything he can to get you to buy."

That critique directly addresses the very pragmatic reason for businesses to avoid deceptive patterns. We'll look at these patterns and how they affect privacy in more detail in Chapter 6, "Avoid Deceptive Patterns," as well as some ways to design using more privacy-friendly versions of these patterns.

However, something as simple as requiring users to disclose their gender or title or to provide an email address or phone number when it seems unnecessary can prompt users to abandon your experience if they perceive that surrendering that information provides them with little benefit. A 2019 survey by Ping Identity, for example, found that 40 percent of respondents had abandoned signing up for an online

---

3 Stephanie Chevalier, "U.S. Shopping Cart Abandonment Rates of Selected Categories 2018," Statista, 13 February 2024, www.statista.com/statistics/232285/most-common-products-services-abandoned-digital-carts-internet-users

4 Rudy Klobas, "Dark Patterns: The Ultimate Conversion Blocker for Ecommerce Websites," The Good, 15 June 2021, https://thegood.com/insights/dark-pattern-ecommerce-ux-design

account because they concluded that a company was demanding too much personal information from them.[5] The same study found that a whopping 89 percent of Gen Zs took advantage of profile settings upon creating an account to lock down their personal data, too.

Loss of User Base

Companies may not just lose users in the moment, though. If data breaches or sloppy treatment of consumer's personal information prove significant enough, they may lead to losses to a company's user base and, in turn, loss of profits and future business opportunities.

For example, a 2017 Cisco report found that businesses experience measurable damage to their business after security breaches.[6] Among their findings:

- 50 percent of companies faced increased public scrutiny after a breach.
- 22 percent lost customers with 40 percent of them losing more than 20 percent of their customer base.
- 29 percent lost revenue, with 38 percent of that group losing more than 20 percent of their revenue.
- 23 percent lost business opportunities, with 42 percent of them losing more than 20 percent of projected future opportunities.

Aside from data breaches, however, intentional changes to an experience that users believe might undermine their privacy and expose them to harm can prompt an exodus from platforms, too.

Since Elon Musk forked over $44 billion to buy Twitter in 2022, he has made a slew of unilateral changes to the platform that have eaten away at its user base, including updates that affect users' online privacy. For example, in late 2024, he announced that users would no longer be able to block one another on the platform now called

---

5 "Ping Identity Survey Finds Greater Appetite for Password Alternatives That Make Login Easy and Prioritize Privacy," Businesswire, 21 September 2021, www.businesswire.com/news/home/20210921005197/en/Ping-Identity-Survey-Finds-Greater-Appetite-for-Password-Alternatives-That-Make-Login-Easy-and-Prioritize-Privacy

6 "Cisco 2017 Annual Cybersecurity Report: Chief Security Officers Reveal True Cost of Breaches and the Actions Organizations Are Taking," Cisco, 31 January 2017, https://newsroom.cisco.com/c/r/newsroom/en/us/a/y2017/m01/cisco-2017-annual-cybersecurity-report-chief-security-officers-reveal-true-cost-of-breaches-and-the-actions-organizations-are-taking.html

X, but that the former blocking feature would become more of a mute function. For just one example, that meant that women, who had blocked trolls for their misogynistic harassment and dogpiling would now be visible to those they felt they had been abused by. (See Figures 4.1, 4.2.) Shortly after, he also announced that X would update its terms of service to use any content users posted there to train X's burgeoning AI feature, Grok. This change triggered another massive exodus that this time included many celebrity, corporate, and other high-profile accounts. Many who left also commented on the platform's increasing toxicity.

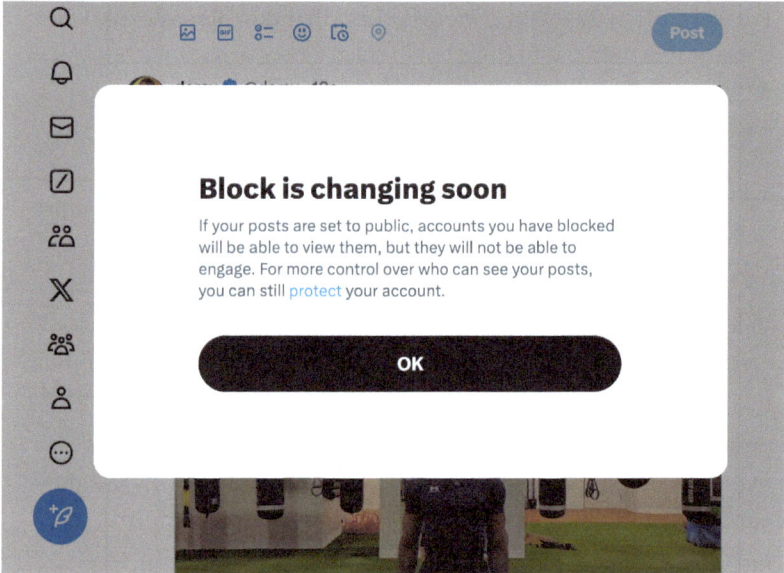

FIGURE 4.1

A module on Elon Musk's X (formerly Twitter) announced that the platform's block function would be "changing." More accurately, it would be removed. The only other option for users would be to lock their accounts, so that no one could see them unless they followed the user.

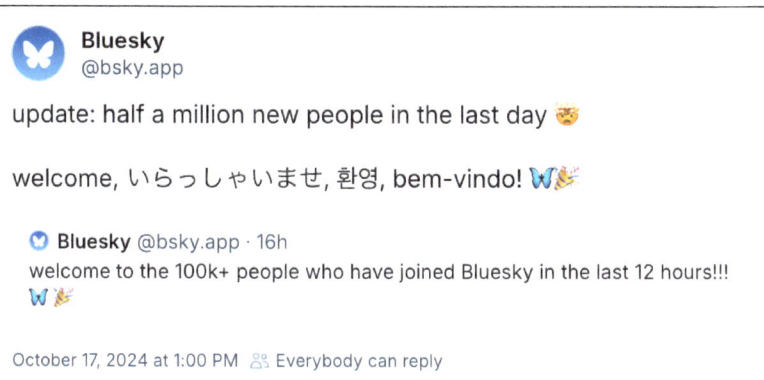

FIGURE 4.2
Twitter rival Bluesky announced half a million new sign-ups within a day after X posted about their plans to remove the block function.

With each update like this, you could see a measurable stream of people leaving the platform to sign up for rival social media platforms like Bluesky, Mastodon, and Threads. Bluesky's developers reportedly even dubbed these moments "EMEs" or "Elon Musk Events."[7] The same day that X announced the change to its block feature, Bluesky rather cheekily tweeted a promotion on Musk's platform: "At Bluesky, we take online safety seriously. If you want to block someone, you can!"

Companies must consider, too, the impact of increased user churn and the difficulty getting traction with maintaining relationships with users, before making changes that may alienate them and drive them away.

---

7 Joshua J. Friedman, Bluesky, 17 October 2024, https://bsky.app/profile/joshuajfriedman.com/post/3l6opxixk2c2d

Penalties and Fines

Last, but certainly not least from a business perspective, consider the aforementioned wallet. Businesses should note the increasing number of penalties and fines that companies have been ordered to pay due to their failure to adhere to emerging privacy laws and regulations. Some companies likely just consider these payments as the cost of doing business, but the fines can be hefty.

For example, in 2021, Norway fined the dating app Grindr a record 65 million kroner ($7.16 million) after the company shared users' data without permission with hundreds of potential advertisers, violating the European Union's General Data Protection Regulation (GDPR). That was the highest fine the Norwegian Data Protection Authority had ordered at that time.

Much larger fines have been levied elsewhere around the globe, however. Some examples are the following:

- In 2021, Luxembourg National Commission for Data Protection (CNDP) hit Amazon with a $888 million fine after a consumer complaint signed by 10,000 people unearthed the fact that the behemoth company was sharing consumers' data with third parties to enable targeted advertising without their consent.
- In 2022, the Irish Data Protection Commission (DPC) fined Instagram $403 million for failing to meet the standards set out in the GDPR for handling the data for children between the ages of 13 to 17.
- The same year, DiDi Global, a Chinese ridesharing app paid a whopping $1.9 billion for breaches of privacy and security laws. To date, that's the largest single fine ever levied against a company for privacy-related infringements.
- In May 2025, the Irish Data Protection Commission fined TikTok $602 million for sending European users' data to China, highlighting, too, that TikTok couldn't guarantee the data wasn't subject to surveillance by the Chinese authorities.

Nonetheless, Meta, the Facebook parent company, takes the grand prize for cumulative privacy-related payouts, having received seven of the ten largest fines for GDPR breaches as of mid-2023.[8] In addition to that Instagram fine, for example, Ireland's DPC later fined Meta a

---

8 Florian Zandt, "Big Tech, Big Fines," Statista, 23 May 2023, www.statista.com/chart/25691/highest-fines-for-gdpr-breaches

further $1.3 billion for GDPR breaches in May 2023. Then, in January 2024, the DPC again levied an over $400 million fine against Meta for forcing users to accept personalized ads to access Facebook. And in September 2024, the DPC fined Meta yet again, this time $1.2 million for storing users' credentials in plain text, which could be accessed by some 20,000 Facebook employees. Two months later, South Korea fined Meta $15 million for illegally supplying advertisers with users' personal data. That totals well over $2.1 billion in fines Meta has paid for just those five settlements.

Although no fines were levied against any companies, the Schrems I and II cases in the EU also had a significant effect upon data privacy for European users. Max Schrems, an Austrian privacy advocate, argued that the Safe Harbor agreement between the U.S. and the EU in 2000 did not adequately protect the data of EU residents. He argued that when data was sent from the EU back to U.S. companies like Facebook, the U.S. government's surveillance practices did not allow for adequate privacy for individuals under EU law and that Europeans had no legal recourse if their data was mishandled by the United States. In 2015, the court agreed and invalidated the Safe Harbor arrangement. Further, in the 2020 follow-up case, Schrems II, the court invalidated the Privacy Shield Framework, which was intended to replace Safe Harbor. These changes have made it more difficult—but not impossible—for Facebook and other companies to transfer EU users' data back to the United States. They can utilize Standard Contractual Clauses with stricter safeguards in place to continue with data transfers.

Additionally, in the United States, despite the lack of robust federal privacy laws, the Federal Trade Commission is cracking down on companies for privacy violations. In 2019, the FTC leveraged a historic $5 billion fine against Facebook for privacy violations, including collecting personal data about children, such as their telephone numbers, without parental knowledge or consent. The FTC added restrictions to Facebook's business operations and reporting requirements to the fine, as well, describing their ruling as "one of the largest penalties ever assessed by the U.S. government for any violation." Similarly, in 2023, the FTC fined Microsoft $20 million dollars for violating child privacy laws. And, in early 2024, the FTC ordered X-Mode (now Outlogic) to stop selling their customer's raw location data without obtaining consent from them first. The FTC also ordered X-Mode to destroy any location data obtained by the company before consumers consented to their collecting it.

Government regulations aside, companies, which ignore privacy concerns can leave themselves open to class action lawsuits, too. No doubt, there's an emerging cottage industry among law firms identifying privacy issues in order to pinpoint and then sue guilty companies, much as we have seen companies sued for the accessibility failures. In 2023, for example, Google lost a class action lawsuit within the United States and was ordered to pay $5 billion for tracking their users when they were using the Chrome browser's "private mode."

These examples are intended to be illustrative, but certainly not exhaustive.

Note, too, that these are generally large companies that often benefit from a global reach. You may need to remind your clients then: The larger the organization, the more likely it is to be targeted with scrutiny for privacy violations.

The Takeaway

Discussing privacy issues with clients or stakeholders can be sticky. You'll need to approach these discussions with tact and diplomacy. But even if there's an up-front cost to designing for privacy, the long-term costs can be devastating to a company's brand and business. You can play a pivotal role in helping your company or clients avoid these penalties and in developing a reputation instead for acting seriously and soberly in attending to users' privacy.

If we could reduce all these points to a single, succinct mantra? Ignoring privacy issues for short-term benefits can lead to greater long-term costs.

CHAPTER 5

Handle Data Responsibly

Only Ask for the Data You *Really* Need	49
Always Obtain Explicit Consent	61
Maintain Data Transparency	63
Make Sure Users Can Delete Their Data	78
The Takeaway	86

When considering how best to maintain people's privacy, your first concern must be how to handle their data conscientiously. Even a single data point misused or accidentally divulged can create a cause for concern. But if you avoid asking for unnecessary data points in the first place, you'll reduce the likelihood of running into unforeseen privacy and security concerns around that data. Remember: Different data points can have different impacts upon a particular individual's privacy. While securing permission to use people's data, you also owe it them to explain how you're using their data and to provide them with ways to maintain or even delete their data.

How seriously should you take your thinking around what data to handle?

In some cases, companies may even decide not to do business with specific users due to factors, such as their age or location and the complications entailed in handling their data. In 2024, for example, many U.S. states began to require age registration to access "adult"- oriented websites. As a direct result, Pornhub locked people in 17 states out of their site, specifically because they didn't want to be responsible for handling users' biometric data (photos of their faces) as those states required. So, yes, a porn site quit doing business with a good portion of the United States over privacy concerns—professedly. (Predictably, demand for VPN [virtual private network] services subsequently spiked in those states.)

Needless to say, online privacy requires handling data responsibly. As a designer, you can play a key role in ensuring that happens. You can make sure that your experience obtains only that data, which is needed, that it's obtained with consent, that the experience is transparent about what's happening with that data, and that you allow users to update or delete their data.

To ensure that you're handling people's data responsibly stick to the following best practices:

- Only ask for the data you need.
- Always obtain explicit consent for that data.
- Confirm that consent is limited to specific purposes.
- Ensure that consent can be withdrawn, too.
- Maintain data transparency.
- Make sure that users can delete their data (or their account).

Only Ask for the Data You *Really* Need

To ensure they remain respectful of people's personal information, companies should practice what's referred to as *data minimization*. That means data controllers should collect, store, and share only the data that's absolutely necessary. That's not just a helpful best practice. In the EU, data minimization is encoded into the GDPR, particularly in Article 5(1)(c), which says data must be "adequate, relevant and limited to what is necessary in relation to the purposes for which they are processed (data minimization).[1]

Companies often ask for data points they arguably don't need. They may use this information for marketing purposes and/or sell it to third parties for profit. Understandably, however, people can be reluctant to hand over certain information that feels deeply personal to them.

Similarly, companies should only use the data they've requested for its expected purpose. For example, if you agree to provide your data to a company for a specific reason, the company might then use that data in concert with data points aggregated elsewhere to draw additional insights you're never made aware of. The GDPR explicitly highlights this "purpose limitation" as a core data protection principle, also adding the principle of "storage limitation," meaning that data should only be stored for as long as it's needed for a specific purpose.

Questions to Ask First

The goal here is to enable you to grow confident in questioning whether the business needs each element of an individual's personal information. In practice, that leads to a series of specific questions for you to ask the business and then yourself as a designer—before you begin designing for an experience. That means completing a privacy check on whether every single element in a form is necessary. In doing so, you'll determine both whether a data point should be requested at all and then whether it can be optional or if it's truly required.

---

1 "Art. 5 GDPR, Principles Relating to Processing of Personal Data," General Data Protection Regulation (GDPR), https://gdpr.eu/article-5-how-to-process-personal-data

Consider these questions to determine what data points to request or to make mandatory:

Questions to ask the business:
- What's the minimum amount of personal data you can ask for?
- Who needs these specific items of personal information within the business?
- Why are they being requested?
- Must the information be required or can it be labeled as optional?
- How will this information be used?
- Who will it be shared with? And why?
- What legal restrictions do you need to consider when requesting this information?

Questions to ask yourself and your team:
- How can you explain to people why this information is being requested?
- How can you offer inclusive options for people to select from when requesting their data?
- If this data is being shared with anyone else, how can you explain that?
- Can you explain all of this honestly and transparently?
- How prominent will that explanation need to be?
- What features do people need to offer or withdraw their consent and to control the use of this data?

You can also develop a more formalized "question protocol" to apply to every form field you include anywhere in your experience as Caroline Jarrett and Gerry Gaffney suggest in their book *Forms That Work: Designing Web Forms for Usability*. They explain that every question you include comes at a cost. There's the literal cost to develop the form field and store the resulting data. And there's the additional cost of potentially losing users if they feel that any questions are unwarranted or that there are just too many of them.

The Value of Someone's PII

Users will likely be concerned any time they're asked to surrender their personally identifiable information (PII). (See Figure 5.1.) It's often but not always characteristic of PII that a single element of it can quickly pinpoint your identity. These data points include your name, email address, phone number, Social Security number (or other government-issued ID number), mother's maiden name, or various types of biometric data.

FIGURE 5.1
A few examples of potential personally identifiable information (PII).

However, bad actors don't necessarily need any elements of PII to identify a single individual. One study concluded that 87 percent of the U.S. population could be pinpointed as specific individuals just by securing their date of birth, gender, and ZIP code.[2] Since they're not unique identifiers, those items aren't typically considered to be PII, so imagine how much harm a criminal could inflict if someone seized upon just three pieces of your actual PII. By latching onto even

---

2 "Personally Identifiable Information (PII)," Imperva, www.imperva.com/learn/data-security/personally-identifiable-information-pii

a couple of these data points, criminals can steal someone's identity. The more personal data someone can harvest about you, the greater the opportunity they have to do you *considerable* harm. Further, cleaning up the mess created by identity theft can take many years. That's why people have a right to know what elements of their PII an entity is collecting and why and who that data is being shared with. And it's also why any experience you design must provide them with an explicit moment to consent to sharing that personal data. Make sure that people feel safe handing over this sensitive information.

For example, when signing up for a new Coinbase account, the cryptocurrency exchange platform asks for the last four digits of your Social Security number. That's an incredibly sensitive piece of PII. Coinbase does provide an explanation for why this data point is needed, however (see Figure 5.2), and the fact that only four digits are required may make potential users more comfortable surrendering part of their SSN, rather than the whole.

Similarly, Coinbase asks for people's citizenship upon sign-up, explaining that this data point is required by law and presenting a link to review more information.

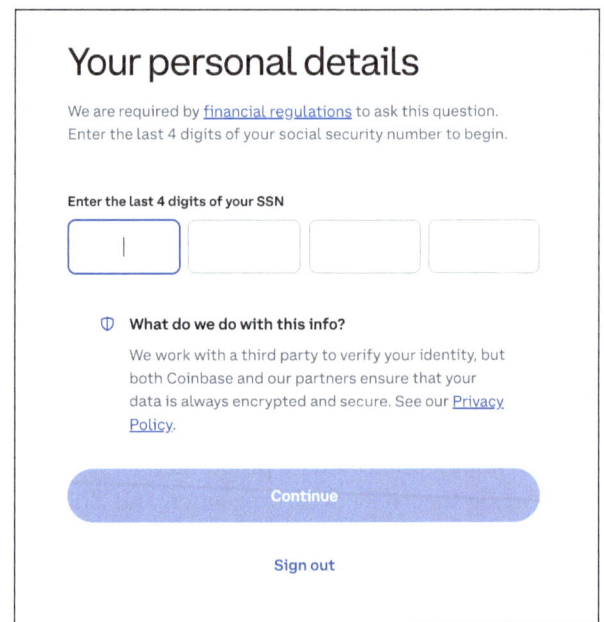

FIGURE 5.2
Coinbase requests the last four digits of a user's Social Security number and presents links to both the financial regulations, which require this information, and their privacy policy, while asserting that users' data will remain encrypted and secure.

MOSAIC THEORY

Jaap-Henk Hoepman references "mosaic theory" in his book *Privacy Is Hard and Seven Other Myths*, an excellent introduction to privacy issues.[3] He explains how all these individual data points people share add up to form a very complex, nuanced portrait of themselves.

"Mosaic theory" featured, tacitly, in the 2012 Supreme Court case United States v. Jones, where the court concluded that if the police place a GPS tracking device on a vehicle, that should be considered a search under the Fourth Amendment, which forbids unreasonable search and seizure. This tracking constituted a privacy issue because the use of such a device undermines the Constitutional concept of a reasonable expectation of privacy. All nine judges concluded that this GPS tracking did indeed constitute a search.

In his concurring opinion, Justice Samuel Alito neatly described what a lower court had called "mosaic theory" without explicitly referencing it:

> Prolonged surveillance reveals types of information not revealed by short-term surveillance, such as what a person does repeatedly, what he does not do, and what he does ensemble. These types of information can each reveal more about a person than does any individual trip viewed in isolation. [...] A person who knows all of another's travels can deduce whether he is a weekly church goer, a heavy drinker, a regular at the gym, an unfaithful husband, an outpatient receiving medical treatment, an associate of particular individuals or political groups—and not just one such fact about a person, but all such facts.[4]

Critics agree that the implementation of mosaic theory could expand the impact of the Fourth Amendment, though they disagree as to whether such an expansion would be practical or not, given the speed with which technology is evolving.

Similarly, in "A Taxonomy of Privacy," law professor and privacy expert Daniel Solove concluded that "When analyzed, aggregated information can reveal new facts about a person that she did not expect would be known about her when the original, isolated data was collected."[5]

continues

---

3   Jaap-Henk Hoepman, *Privacy Is Hard and Seven Other Myths* (The MIT Press, 2021).

4   *United States v. Jones*, U.S. 565 (2012)

5   Daniel Solove, "A Taxonomy of Privacy," *University of Pennsylvania Law Review* 154, 2006: 506, https://scholarship.law.upenn.edu/penn_law_review/vol154/iss3/1

> **MOSAIC THEORY** (continued)
>
> In *Doppelganger* Naomi Klein impresses upon us the fact that we're each creating a sort of digital doppelganger online, as we leave a trail of data in our wake wherever we go. These digital golems may not even be an accurate, exact representation of ourselves, but, even if unfairly cartoonish, they're still enough for companies or other government entities to know your wants, your preferences, your desires, even your physical location. You may be able to delete parts of your doppelganger, modify it to degrees, but, unless you're unusually protective of your identity, this shadow self remains.
>
> Consequently, when people conclude that surrendering a single element of personal data may not seem intolerable or threatening, they should remember that it's the gradual accretion of these details that proves troubling. It's tempting to reference the old metaphor of frogs boiling in water here to describe how we have all become acclimatized to sharing our data until it's too late. The problem, however? The frogs boiling in water metaphor isn't true. Turn up the heat and the frogs will jump out. So far, we have not.

Handle Other Personal Data Sensitively, Too

It's not just PII that you need to handle carefully. More broadly, when personal data isn't requested in a thoughtful way, people may abandon an experience rather than provide it or find it's impossible to proceed even if they would like to. Or they may just enter some fake information to proceed. Considering these possibilities means thinking inclusively, a practice which ensures that you're designing better experiences anyway. (More on that in Chapter 9, "Cultivating a Culture for Privacy by Design.")

Consider, for example, why do you need to know somebody's gender when they're completing a form? Why do you need to know their title? Title is often used to infer gender, when an organization understands it may seem impolite to ask for the latter. If these data points are truly needed, be prepared to explain *why* they're needed at the point where you're asking for them.

In her article, "How to Ask About Gender in Forms Respectfully," Ruth Dillon-Mansfield shares more inclusive patterns for asking for gender and title—if they're truly needed.[6] (See Figure 5.3.) She suggests a Golden Rule for requesting personal information, too: "What are we using the information for? Would *not* collecting it cause a problem? Will people feel comfortable answering?"

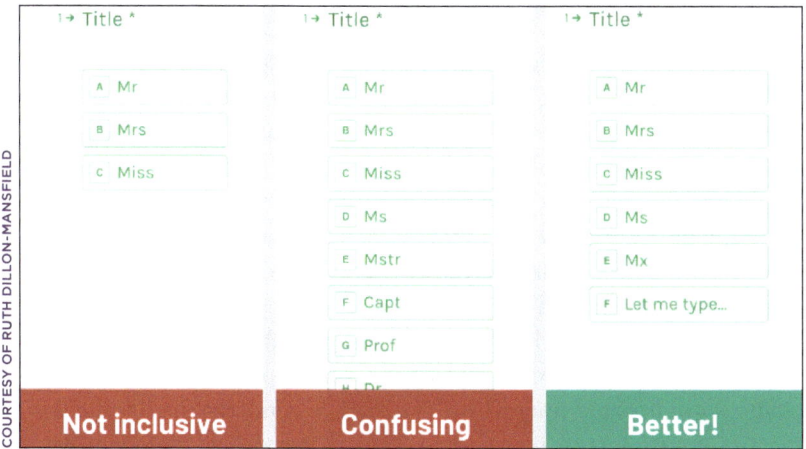

FIGURE 5.3
Ruth Dillon-Mansfield's "How to Ask About Gender in Forms Respectfully" shows two less helpful ways to ask for someone's title, plus a better direction.

She concludes, "If we ask ourselves why we're requesting a particular piece of information from someone, and nobody can answer, it's probably a good sign we shouldn't be asking at all."

Hamstrung by Titles

A UK couple feared they might lose their mortgage simply because their bank didn't allow a nonbinary title in their application, explaining that it couldn't be accepted in their system. G Sabini-Roberts is a branding consultant, who is nonbinary and uses they/them pronouns. They were unable to complete the mortgage form because they use the title Mx., a title that has been a legal option in the UK since 2011. Thankfully, the situation didn't escalate. Sabini-Roberts and the bank had a polite conversation, and the bank proved helpful, enabling the couple to proceed with their mortgage.

---

6 Ruth Dillon-Mansfield, "How to Ask About Gender in Forms Respectfully," Ruth Dillon-Mansfield, https://ruth-dm.co.uk/posts/how-to-ask-about-gender-in-forms-respectfully

As G told *Metro*, they thought, "'I can't be their first trans nonbinary customer.' I was scared. This might mean we lose the house, which we've already invested so much in."[7] G explained that this story was often mistakenly reported as an issue with pronouns. But, as they explained on Facebook, "This was always about simply getting a legal document in my legal name."[8]

Still, from a privacy by design perspective, your takeaway from this incident would be to ask: Why did the bank need a title at all? Maybe there is some particular reason that banks need titles to accompany their customers' personal information in the 21st century. If so, they should be prepared to explain why, or they should just drop the requirement—especially if it could be construed as presenting a hurdle for some people in the consideration of their being qualified for a financial institution's products and services. In this case, the problem was two-fold: The bank was asking for the customer's title, which could be considered a privacy issue as it potentially divulges gender without explaining why it's needed. Second, the form made the mistake of not being inclusive around the customer's gender identity.

This proves to be a remarkably common issue. As of this writing, Mx. Sabini-Roberts would not be able to buy a plane ticket on a popular airline's website. Let's look at how JetBlue currently handles form fields, which require you to enter personal information.

The first field you're required to complete on the form to provide your "Traveler Details" is "Title." There's no explanation why. (See Figure 5.4.)

They do provide a long list of titles in the drop-down, so you're in luck if you're a Captain, Baron, or Viscountess. Not so much if you're a Mx. Nor do you have the option to add any other title.

Somewhat helpfully, when asking for gender, JetBlue provides an explanation via tooltip that you can "book by selecting any of the available options, as long as the name and date of birth you provide matches your government-issued ID." (See Figure 5.5.)

---

7   Emma Brazell, "Couple Refused Mortgage Because Bank Didn't Recognise Non-binary Title," *Metro*, 8 February 2022, https://metro.co.uk/2022/02/08/couple-refused-mortgage-because-halifax-didnt-recognise-non-binary-title-16070764/

8   G Sabini-Roberts, Facebook (post), 7 February 2022, www.facebook.com/share/p/1KmLWj5zni

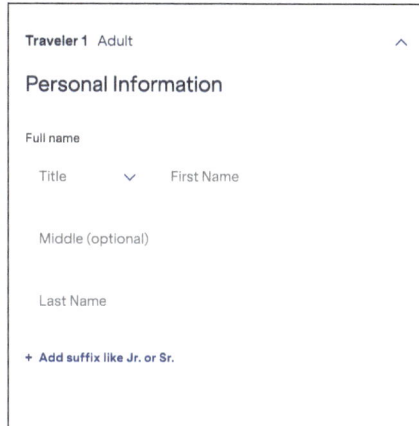
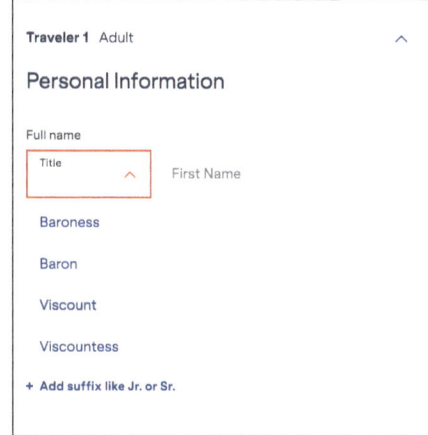

FIGURE 5.4

The JetBlue site shows Title as a required field to continue booking a flight. The activated drop-down shows some of the idiosyncratic—but still not inclusive—titles JetBlue includes in the drop-down.

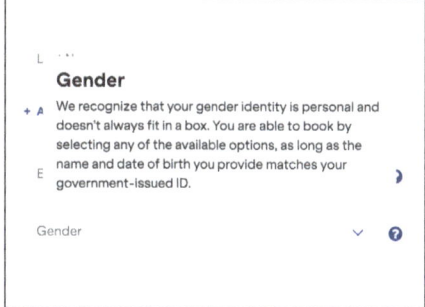
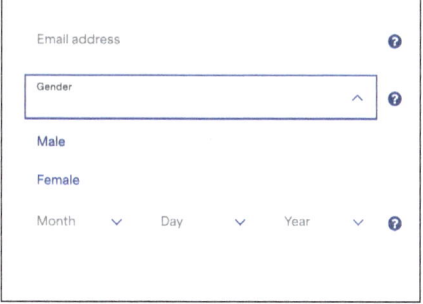

FIGURE 5.5

The JetBlue site indicates that gender is a required field, along with a tooltip, ostensibly explaining why. The activated drop-down indicates only two options for gender: Male and Female.

The available options within the drop-down are "female" or "male." Unfortunately, this ignores the fact that some government IDs do allow for a "nonbinary" selection or even "no gender." Further, in the United States you are able to use either of those gender markers on your passport, a fact that seemed to elude the U.S. airline. At the time of this writing, however, the Trump administration announced via an executive order on gender that people may only use male or female to describe their gender on new U.S. passports going forward.

Nonetheless, existing passports may still include those markers. (This limitation on how people must describe their gender is arguably a violation of privacy itself.)

I asked G whether they would be able to purchase a ticket on JetBlue's site. They confirmed they could not.

JetBlue isn't alone in this issue, however. Although they don't request gender, Spirit requires a title and only provides the selections of Mr., Mrs., and Ms. Currently, American, Delta, United, and Alaska Airlines do provide more thoughtful and inclusive experiences for nonbinary people. They don't require a title, and although they do require customers to provide their gender, they allow the additional helpful options of "Undisclosed" and "Unspecified."

New Contexts, New Meaning

Remember, too, that sometimes certain data points that previously seemed innocuous can suddenly take on new meaning. After the Supreme Court overturned Roe v. Wade, for example, some women shared on social media that they had deleted period trackers, when those apps began requiring them to share the state they lived in to use them. In April 2024, a Redditor with the handle BabserellaWT posted a thread headlined "My ovulation tracker suddenly asked what state I live in." "I deleted it immediately," she wrote. "Like I'm gonna risk police showing up to my door if I miss a month due to nonpregnancy reasons."[9]

Note the screenshots from the app she deleted, Ovia. (See Figure 5.6) To create an account, you're required to provide your state of residence. A tooltip explains, "We ask for your state so we can comply with the privacy laws that apply to you."

However, if your focus as a company is on privacy by design, you would first ask, "Why not offer the most private experience we can conceive of for all users, regardless of which state they live in?" The added bonus? Such a company would also more likely be compliant with policies in other countries and regions like the EU, too. They may, however, make less money, which is obviously a motivator for many companies to *not* provide the most privacy-centric experience imaginable.

---

9 BabserellaWT, Reddit (post), April 2024, www.reddit.com/r/WitchesVsPatriarchy /comments/1btfoj9/my_ovulation_tracker_suddenly_asked_what_state_i

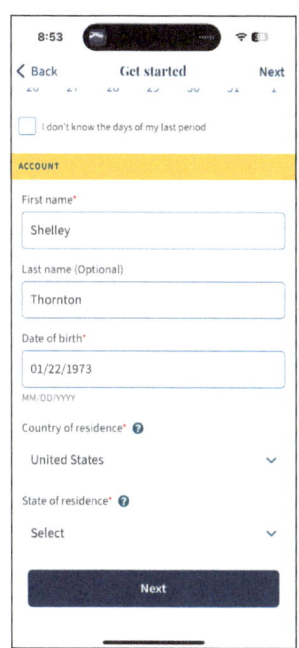 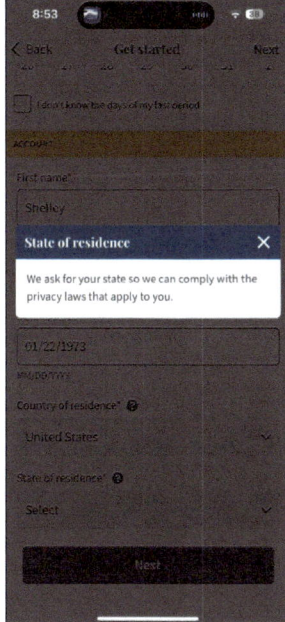

FIGURE 5.6
The Ovia app shows that a state of residence is required, but the tooltip copy only explains that the company is complying with privacy laws. It does not address any specific privacy concerns.

Clearly, some U.S. companies fear they'll have to respond to government demands for user data. However, they can refuse to fulfill subpoenas for that data. In fact, one such app, Period Tracker, promised to do just that, saying "We would rather close down the company than be an accomplice to this type of government overreach and privacy violation."[10] Still, many companies would cooperate with a government entity in these situations to avoid any penalties. In those moments, they'll be taking a side. And it won't be the side of their users' privacy.

These examples highlight just a few common elements of personal information that companies often require people to submit, but there are many more. If you do want to obtain certain data points to, say, encourage users to complete a profile, consider at least making as many of those elements—photos, addresses, age, gender, and so on—as optional as possible.

---

10 "On Data Privacy," GP Apps, 19 May 2022, https://gpapps.com/2022/05/19/on-data-privacy

REAL NAMES AND THE NEED FOR ANONYMITY

Because of the sheer volume of online abuse some companies have to deal with, especially social media platforms, they often consider implementing a "real name" policy. While this practice might initially sound like a commonsense solution, in practice, it often causes more harm than good, since it fails to consider the many legitimate reasons that good faith users may have for maintaining their anonymity, especially those within various at-risk communities. The use cases for remaining anonymous can vary tremendously: A woman who has been stalked by her former spouse. A human rights advocate living under a repressive regime. An LGTBQIA+ teen exploring their identity while living with abusive parents. Think also of sex workers, journalists, and whistleblowers in different situations. Removing the shield of anonymity could be devastating for individuals in these situations.

If you're not part of the LGTBQIA+ community, for example, you may not realize just how many young people depend on anonymity to reach out for comfort, community, and helpful, sometimes life-saving information. Previously, I volunteered with a nonprofit organization that worked with current and former students at a fundamentalist Christian university located in South Carolina. If discovered, the LGTBQIA+ students enrolled there could face not only expulsion from the school, but also getting kicked out of their family homes and ostracized from their churches and families. They provide an excellent example of why some people may need to adopt an anonymous identity online.

Obviously, using some platforms will require a real name, but those social platforms that allow people to interact with one another in large numbers, whether in public or even privately, do need to consider the very legitimate use cases for remaining anonymous.

Companies like Apple, Google, and Microsoft do offer "identity as a service" features that would enable people to access other sites and prove their identity without sharing specific data points, such as their name. That would allow for anonymity, though it also requires the user's ongoing trust in the company providing the service from a security perspective.

Always Obtain Explicit Consent

Have you noticed how many times the word *consent* has cropped up in your reading already? That's because consent is a singularly important concept in the field of privacy and the practice of privacy by design. Indeed, Dr. Cavoukian mentions the word five times within her relatively short manifesto on privacy by design, and she lists consent first when articulating four primary fair information practices or "FIPs": Consent, Accuracy, Access, and Compliance.[11]

Consent, by definition, must never be *presumed*. It must always be explicitly acknowledged. You can't trick people into consenting either. If you're tempted to trick people into providing data, take a moment to consider why that temptation arose. Does it mean you don't believe you have a justifiable reason for requesting that data? Assuming you can justify the request, you should still take time to explain to people why you need their data, so they can feel confident in signaling their consent.

Additionally, ensure that users understand what their consent applies to and what it doesn't. That means explicitly describing what purposes the consent is needed for and confirming that the user's consent is limited to those purposes.

Planned Parenthood came up with a way to explain consent in the context of sex that you can keep in mind in the context of privacy by design, too. They use the acronym FRIES, meaning that consent should be Freely given, Reversible, Informed, Enthusiastic, and Specific.[12]

"Reversible" is an important characteristic there as it indicates that for consent to be authentic, people must have the ability to withdraw it. That means that any experience you're developing must include features that enable users to withdraw their consent, too.

Finally, consider the concept of "consent fatigue"—something we all likely suffer from. Users literally begin to feel fatigue and indifference over having to surrender control of their data so often. Some companies take advantage of this fatigue to secure users' data. Companies hoping to practice privacy by design will respect your tendency toward consent fatigue and ensure that you feel safe and informed when handing over your data.

---

11 Ann Cavoukian, Ph.D, "Privacy by Design: The 7 Foundational Principles: Implementation and Mapping of Fair Information Practices," Report, 2010, Information and Privacy Commissioner of Ontario, Ontario, Canada.

12 "Sexual Consent," Planned Parenthood, www.plannedparenthood.org/learn/relationships/sexual-consent

MARK ZUCKERBERG AND "RADICAL TRANSPARENCY"

"For better or worse, Facebook is an incredibly important platform for civil life, but the company is not optimized for civil life. It is optimized for hoovering data and making profits."—Ethan Zuckerman

Facebook's founder Mark Zuckerberg once famously said he believed in the principle of "radical transparency," an idea that may have some merit in specific contexts, such as when companies provide clear insight into how they're utilizing your data and who they're sharing it with. Unfortunately, Zuckerberg has often made decisions that have affected users' privacy, including famously changing Facebook's settings to make users' likes and other preferences public without their consent. Clearly, he hopes to increase engagement on the platform and, in turn revenue. Nonetheless, given his perceived flippancy around privacy, Zuckerberg has repeatedly found himself on the receiving end of criticism from experts in the privacy community, as well as the broader human rights community.

If transparency is good, it still must be allowed via consent—especially when contemplating people's personal information—and not enforced by default. It's ironic then that companies like Meta, Facebook's parent company, fail to be transparent with how they handle people's data, but expect transparency by default from their users. The counter to that potentially invasive directive could be practicing, or at least aiming for, radical empathy. It provides a better pathway for developing successful and thoughtful user-centered experiences than radical transparency, which can fall prey to *assuming* consent.

In some places like the United States, assuming consent is certainly legal, and, arguably, that assumption *could* be handled respectfully. Assumed consent may prove less harmful in some situations than others. Consider the difference between sharing user's anonymized browsing data with third parties versus suddenly making someone's sexual orientation public. Either way, however, assuming consent still fails to meet the principles of privacy by design as described in Dr. Cavoukian's privacy by design framework: Privacy should always be the default setting.

As far as radical transparency goes, the degree to which an individual's personal information becomes publicly available should remain under the control of that individual, not a corporation or organization that may suddenly find itself eager, for example, to display that information to trigger engagement. Refusing to be transparent about how you handle someone else's data will be perceived as an ethical failure, not just a design failure. Similarly, expecting others to be transparent about their personal details without asking them first will be noted as an ethical failure, too, because it's a violation of an individual's privacy.

Maintain Data Transparency

If companies hope to emphasize transparency, an excellent place for them to start would be with clearly explaining their use of people's data. Once you've established that it's acceptable to request someone's personal information and you seek to secure their consent for it, be transparent about exactly how their personal data will be used, especially if it may be shared with third parties.

This hidden sharing may seem like technical, back-end stuff, so how can you assist with this as a designer? Luckily, you're perfectly positioned to advise stakeholders and clients on how to explain to people how their data is being used—as well as to determine how prominently this information can be positioned and accessed. And, hopefully, you're positioned and empowered to establish some thoughtful design patterns, which neatly and effectively communicate this information at just the right moment within an experience.

Sharing Often Exceeds Expectations

Perhaps the greatest area needing improvement in the realm of online privacy is the lack of transparency about exactly how everyone's data is used, and who it is shared with. For example, most people clicking that "accept" button on a cookie banner to dismiss it so they can read a news article likely give little thought to how many third parties their information will be shared with. After all, *cookie* sounds like such an innocent term, doesn't it? Who doesn't love tasty cookies? A term like *tracer* or *tracker* would be more transparent.

You might guess that your information is being shared with several parties. Maybe dozens? You might not suspect the reality: It's not uncommon for these platforms to share your data with hundreds of third-party companies. Remember that 2024 *Wired* study that showed that some of the most popular websites on the planet share their visitors' data with well over 1,000 parties?[13] Each individual Facebook user is tracked by thousands of companies, too.[14]

---

13 Matt Burgess, "Some of the Most Popular Websites Share Your Data with over 1,500 Companies," *Wired*, 21 March 2024, https://wired.me/technology/the-most-popular-websites-share-data-with-over-1500-companies

14 Jon Keegan, "Each Facebook User Is Monitored by Thousands of Companies," The Markup, 17 January 2024, https://themarkup.org/privacy/2024/01/17/each-facebook-user-is-monitored-by-thousands-of-companies-study-indicates

I once studied the cookie banner behavior on a single news site and discovered that they shared their users' browsing behavior with over 600 third parties. Somewhat amusingly, they offered checkboxes next to each party, so you could deselect them one at a time if you wished. It would be fascinating to review the analytics on how many visitors to that paper's site ever dug down far enough to find those settings, let alone to uncheck them, one at a time. It's unlikely the vast, overwhelming number of visitors to the site understood just how many third parties their data was being shared with.

What sort of data is that? Well, that would include not just data about the site you're visiting—let's say a newspaper—but also about your browsing behavior *before* coming to that site and *after* leaving it, as well as your preferences, plus whatever demographic information that platform can glean about you, such as age, gender, and location.

The intimate nature of this data sharing can prove troubling, too. In early 2024, Mozilla announced a study on the proliferation of virtual boyfriends and girlfriends—basically AI chatbots, which promised opportunities for digital romantic interludes.[15] The problem was that most of the companies running these bots sold or shared the content that users dropped into their sessions. Folks romancing these bots might be surprised to learn that this highly personal data could include information about their sexual health, medical prescriptions, or gender-affirming care.

Gizmodo covered this report, describing the 11 apps Mozilla examined as responsible for a "data harvesting horror show."[16] What else did Mozilla discover about these apps, which had, at that time, garnered 100 million downloads on the Google Play Store alone?

Some stats:

- 90% of the apps sold or shared user data for targeted advertising or other purposes.
- 54% of them didn't allow users to delete their data.

---

15 Jen Caltrider, Misha Rykov, and Zoë MacDonald, "Romantic AI Chatbots Don't Have Your Privacy at Heart," Mozilla, 14 February 2024, https://foundation.mozilla.org/en/privacynotincluded/articles/happy-valentines-day-romantic-ai-chatbots-dont-have-your-privacy-at-heart

16 Thomas Germain, "Your AI Girlfriend Is a Data-Harvesting Horror Show," Gizmodo, 14 February 2024, https://gizmodo.com/your-ai-girlfriend-is-a-data-harvesting-horror-show-1851253284

- They deployed an average 2,663 trackers per minute.
- These stats were admittedly skewed by one app, Romantic AI, which somehow managed to set 24,354 trackers loose within a minute of your opening the app.

The breadth of this sort of data collection certainly varies from platform to platform. Nonetheless, it's often at a scale not immediately understood by the users of these experiences.

Some may argue that people have tacitly accepted this data sharing, as the price they pay for access to ostensibly free content on the internet—a sort of handshake they exchange every time they open a browser window. In the case of "clickwrap agreements," users actually enter into a contractual agreement with a company, when they agree with their Terms of Service or Privacy Policy in order to sign up. However, due to the opaque nature of these experiences, people simply do not always grasp how much they're handing over. (Imagine having to pay for your groceries without having any idea how much they cost!)

Circumstances can change overnight, too, creating new situations that give people new and meaningful concerns for their privacy. For example, a company you once trusted could get bought by a company you know little or nothing about, which then abruptly changes who they're sharing your data with. After Elon Musk bought Twitter, the evolving company, now dubbed X, changed its privacy policy to allow for harvesting people's posts to train their AI model Grok. Users could opt out, but they were never given the opportunity to opt in. Later reporting suggested that users' data might be gobbled up for training AI even if they did opt out.

Similarly, people's experiences vary tremendously by geographic location, so they might be comfortable sharing their data broadly in one place, but not so extensively in another.

Will you as a designer have much, if any, influence over the scope of data sharing the companies you're working with indulge in? Or to question or provide feedback on their very business models? Likely not. However, you may have opportunities to highlight and explain that scope and to provide tools that allow users to control the flow of their information to some degree.

OVERSHARING AND UNDEREXPLAINING

Remember the Chinese robotics company Ecovacs from the introduction whose vacuum cleaners were uploading video, audio, and photos of their owners to help train the company's AI?[17] That company was certainly not transparent about the material their roaming robotic cleaner was sharing. Nor were the vehicle manufacturers like Nissan and Kia, which you learned were sharing data around "sexual activity" and your "sex life" from your vehicle—though that fact may have been squirreled away in their privacy statements, technically.[18]

What about the sleep app you might use which listens to you during bedtime hours? Is it transparent about what it's doing with that data? Is it recording you? Is it just storing this information on your phone? Or is it sending it to the company's servers? To any other companies' servers? In fact, many of these app companies share users' intensely personal data with third parties, such as their business and marketing partners. (Worth noting: Some companies stress that this data has been anonymized, but researchers have shown that data can be quickly de-anonymized using remaining demographic data.) Users have a right to know these details, and you have a responsibility to consider how to make this all crystal clear to them. And up front.

---

17 Julian Fell, "Insecure Deebot Robot Vacuums Collect Photos and Audio to Train AI," ABC News, 4 October 2024, www.abc.net.au/news/2024-10-05/robot-vacuum-deebot-ecovacs-photos-ai/104416632

18 Jen Caltrider, Misha Rykov and Zoë MacDonald, "It's Official: Cars Are the Worst Product Category We Have Ever Reviewed for Privacy," Mozilla, 6 September 2023, https://foundation.mozilla.org/en/privacynotincluded/articles/its-official-cars-are-the-worst-product-category-we-have-ever-reviewed-for-privacy

The What, Why, Who, and How of Data Sharing

To be truly transparent about data sharing, it's important for you to be specific about *what* data points you're using and *why*, as well as *who* you're potentially handing that data off to. Additionally, you can build trust with users by explaining *how* their data is collected, stored, traced, and controlled, too.

This practice of data transparency need not just be an onerous box to check off in order to adhere to a best practice, however. Approach this moment as an opportunity to explain the benefits of sharing your users' data, too. (Presuming there are any!)

First, take a moment to ask yourself the same questions your users will be asking:

- Does sharing this data ensure a better experience in the future?
- Does it help personalize advertising or offers for them?
- Is that really enough to justify asking for the information?

If these truly are benefits, be prepared to explain them in detail. And if you can't cobble together a convincing explanation, consider whether you're designing the right product. In some situations, if the data collection could actively cause people harm, you may need to ask, "Should this product exist at all?"

In the following example, Microsoft uses three screens to highlight privacy matters, which you review before you can use Outlook (see Figure 5.7). These screens also include links to privacy information on the specific topics about the use of diagnostic data, your privacy settings, and the ability to examine your work to provide suggestions and recommendations. The second screen also includes a way to opt out of sending optional data about Microsoft's suite of products (not just Outlook). Note that neither of the radio buttons beside these Yes or No options is preselected.

These are all helpful features, though it's worth noting that Microsoft does *not* clarify here how many third parties they may be selling your data to or sharing it with. Additionally, although the title for the first screen reads "Microsoft respects your privacy," the breadth of how much information Microsoft collects isn't clear because there's no summary of how your privacy settings affects that. Then, although the first screen includes some explanation around diagnostic data, the second asks whether the user would like to share diagnostic *and* usage data, the latter being quite a broad category that isn't explained to the user here. Understandably, Microsoft wants to keep these screens as lean and succinct as possible to propel users through onboarding quickly, but this pattern reveals how difficult it can be for users to divine just exactly what it is they're sharing even when companies make some effort to show them. In reality, our decisions often come down to a simple gut feeling: Do I trust this company with my data?

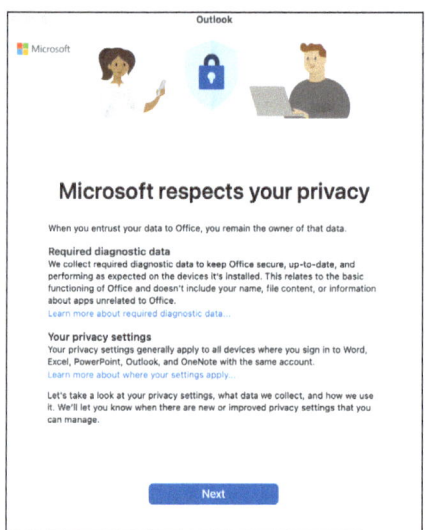
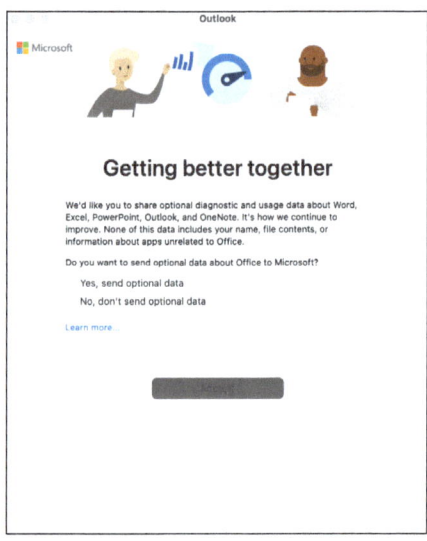

FIGURE 5.7

In these onboarding screens for Outlook, Microsoft reviews the potential use of your data with you, provides links to additional information, and requires you to make a conscious choice to enable them to collect your data, rather than assuming it by default.

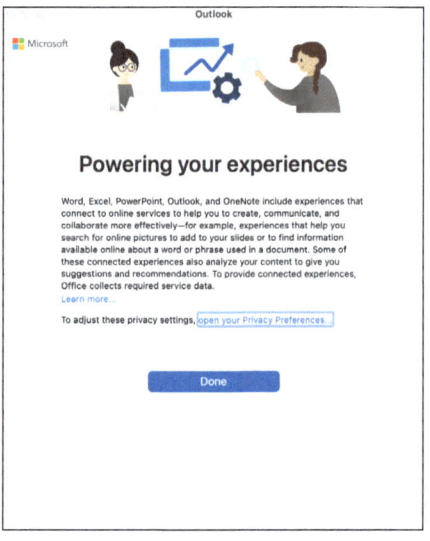

NOTE **THE *WHO* GETS COMPLICATED**

Being transparent about who's going to receive people's data may mean listing out all those third parties involved in the sharing or selling of that data or, at least, the categories they fall into. Many companies don't do that yet, although they may be required to do it in the future, and they are already required to do so under some laws like the GDPR. For sharing data, those categories might include, for example, content management systems, customer management systems, email marketing vendors, and financial transactions processors. For selling data, those categories might include advertising networks, consumer insights companies, data brokers, email marketing providers, marketing agencies, and market research firms among others.

Given that many companies share users' data with data brokers, those companies may have no idea how many additional parties that data is being shared with themselves. That means, currently, the best you can do may be to take a good faith stab at listing out the parties your company shares data with *directly*.

Here are some best practices for clearly explaining the use of users' data:

- Provide clear, straightforward explanation about what data is shared during onboarding and why it's needed. (See Chapter 8, "Provide Tools for Enabling Privacy," for more about the importance of onboarding.)
- State how many third parties the user's data will be shared with.
- Consider linking to a full listing of third parties, at least by category.
- Include an explicit call to action (CTA) to request consent, or explain with some copy, that by proceeding, a user agrees to consent to sharing their data.
- Don't preselect any options on behalf of your users.
- Explain how their data is being protected as the company collects and utilizes it.

LEMONADE SHOWS HOW THEY USE YOUR DATA

Some companies do go to pains to set standards for what data they're using and why. For example, check out the home insurance app Lemonade. The company includes a "Privacy Pledge" link listed under "Company" in their footer. This placement is more prominent than where companies typically place this information—at the *very* bottom of the page in a dedicated line in the smallest font you'll find on the site. Instead, Lemonade has elevated this content to a more significant area, indicating it's something they want you to know about their company.

On that privacy page, Lemonade presents a table of *what* elements of personal data they're collecting from you, as well as an explanation around *why* they believe those data points are needed. (See Figure 5.8.) The table is scannable and remarkably free of jargon. It also addresses the *who* question—which third parties certain types of data may be shared with.

Also, at the top of the page and placed prominently beneath the page title, to help address the who question, Lemonade summarizes their privacy policy with a pledge: "TL;DR: We will never, ever, sell your data to anyone." (See Figure 5.9.) (Note, however, that a company need not sell your data directly to a third party to still profit off the sale of demographic groups or cohorts you're a part of that can drive ad personalization.)

Significantly, the company uses very straightforward language to explain their entire privacy policy. You'll learn more about the importance of language in conveying privacy issues in Chapter 7, "Use Language with Care."

Pet Life Giveback		*Lemonade*		My Account

We may disclose your information to third parties outside of our company. These external recipients may include, in particular:

TYPE OF DATA	COLLECTED AND REQUIRED TO PROVIDE INSURANCE SERVICES	MAY BE SHARED WITH A THIRD PARTY TO IMPROVE OUR SERVICES	MAY BE SHARED WITH A THIRD PARTY FOR CROSS-CONTEXT BEHAVIORAL ADVERTISING
Images, audio and video recordings	✓	✓ Claim support service providers	✕
Bank account number	✓ Claim payments	✓ Secure payment processors (to facilitate claim payments)	✕
Biometric information (such as identity-verifying voice, handwriting and facial features, excluding behavioral characteristics, like driving behaviors, etc.)	✕	✕	✕
Characteristics of protected classifications (e.g., age, sex, race, ethnicity, physical or mental handicap, etc.)	Age, gender ✓	✓ Analytics providers; service providers that assist with the employment applicant hiring process (only in accordance with applicable law)	✕

FIGURE 5.8
A much longer scrolling table from Lemonade shows the types of data they use and don't, as well as why it's used and who it's shared with.

Lemonade

Lemonade's Data Privacy Pledge

TL;DR: We will never, ever, sell your data to anyone.

FIGURE 5.9
Lemonade has presented this prominent privacy pledge on their site for several years now: "We will never, ever **sell** your data to anyone."

Handle Data Responsibly

Provide Clear Patterns for Disclosure

You can educate users about how their data is being used by designing clear, succinct patterns to disclose that information within the experience. Refer to the following best practices for privacy when designing for these moments, which should inform users and cultivate their trust:

- Ensure immediate, prominent ability to deny consent.
- Ensure "Accept" and "Deny" options are equally weighted visually.
- Don't use deceptive language.
- Avoid using preselected checkboxes.
- Link to a privacy policy, any related content, and any more detailed settings.

Currently, cookie banners often provide bad examples of disclosure, because they're designed to get you to proceed as quickly as possible without reviewing what you're agreeing to when you click those innocent little words "Allow" or "Accept." Since many of those cookie banners are designed to trick you, they qualify as "dark" or deceptive patterns. You'll review deceptive patterns in much more detail in Chapter 6, "Avoid Deceptive Patterns."

Cookie Banner 1: Confirmation with Buried Consent

One of the most common patterns for cookie banners allows you to accept the cookies with just one click but forces you to click multiple times to review the full listing of options. (See Figure 5.10.) In other words, it creates friction for creating a more private experience but removes friction to accept a less private one. This pattern has been banned by some regulatory bodies, such as the French Data Protection Authority (CNIL), which insists that the act of rejecting cookies must be as direct as accepting them.[19]

---

19 "Dark Patterns in Cookie Banners: CNIL Issues Formal Notice to Website Publishers," CNIL, 12 December 2024, www.cnil.fr/en/dark-patterns-cookie-banners-cnil-issues-formal-notice-website-publishers

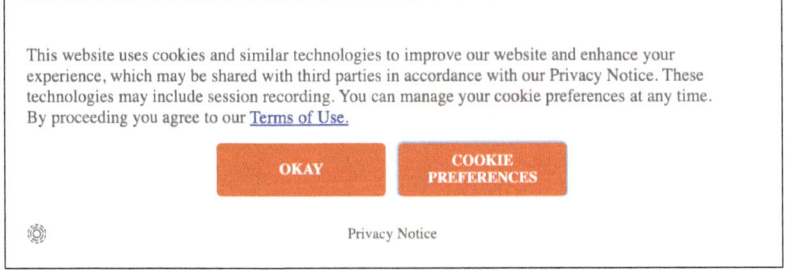

FIGURE 5.10
This banner from the Shutterfly website presents the cookie choices as equal, but the second button requires you to click into a separate screen and deselect checkboxes to deny consent.

Cookie Banner 2: Confirmation Without Consent

Perhaps the worst imaginable design—aside from no option or indication of tracking at all—is one that expects you to proceed only by accepting and offers no way to deny tracking at all. In this example from Digiday, the copy vaguely suggests that you can get an overview of the cookies used by visiting "your personal settings," but no link is provided to these settings (see Figure 5.11). Ironically, the banner accompanied a privacy-related story on this site devoted to covering online media.

This design pattern does not allow for consent: It's really a cookie acknowledgment banner, not a cookie consent banner.

Similarly, this cookie banner from a travel-oriented site uses jokey, informal language—"What's the deal with cookies?"—to prompt users to accept tracking but provides no way for them to opt out. (See Figure 5.12.)

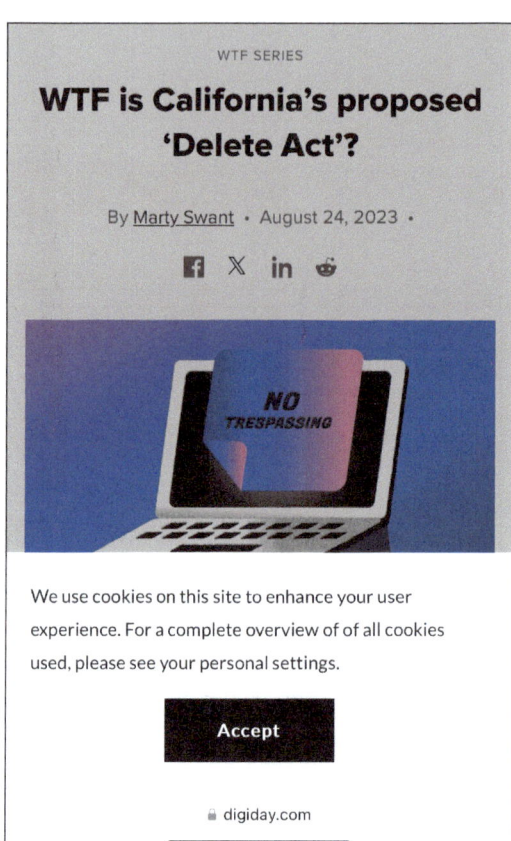

FIGURE 5.11
Digiday's cookie banner offers visitors no option other than to "Accept" cookies. Neither does it include a link to the mentioned personal settings.

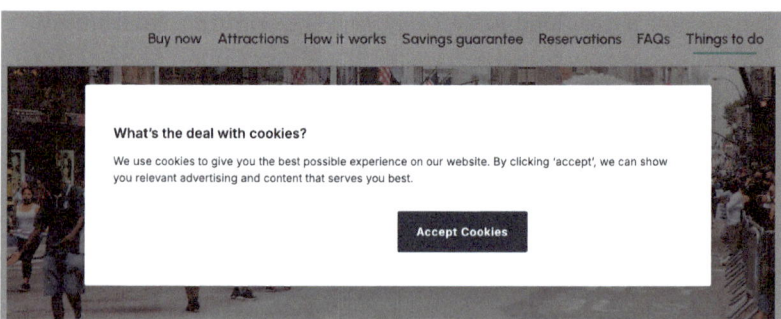

FIGURE 5.12
Visitors to the New York Pass website had no choice but to accept cookies. The updated site now includes the more privacy-friendly options to "Reject All" cookies or "Review Purposes."

Cookie Banner 3: Confirmation with Equal Consent

Note the difference in the banner in Figure 5.13 from the Climate Designers community site that allows you to "Reject Targeting Cookies," providing a more explicit moment for consent. This banner also includes a link to "more information," providing a direct link to the site's privacy policy. That's another best practice for cookie banners.

Even in this case where users do have the option to "Reject Targeting Cookies," the treatment isn't *truly* equal: The "Accept" button is treated visually (green for go!) as the primary option, while the "Reject" button is treated as a secondary one.

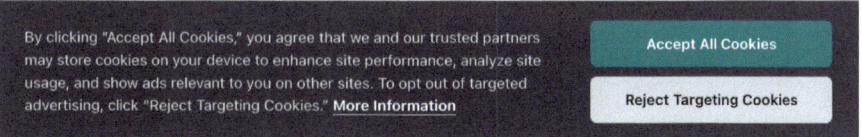

FIGURE 5.13
This banner provides the straightforward ability to either "Accept All Cookies" or to "Reject Targeting Cookies." Still, this wording does allow for other types of cookies.

> **NOTE** THE IMPORTANCE OF LANGUAGE
>
> The preceding example demonstrates the importance of clear language in calls to action, too. Subtle differences in that copy can conceal important distinctions. Here, for example, the "Reject Targeting Cookies" CTA sounds helpful but likely still allows for other types of "non-essential" cookies. See Chapter 7, "Use Language with Care."

Cookie Banner 4: Confirmation with Consent, Additional Options

Here's an example of an equally weighted cookie banner. It's taken from the Cookiebot public website itself, so the company appears to be taking these best practices to heart and allows for them in varying configurations within their banners. (See Figure 5.14.) In fact, Cookiebot promotes their "website cookie consent tool" as providing a path to "effortless website privacy compliance," specifically referencing the GDPR and California's Consumer Privacy Act.

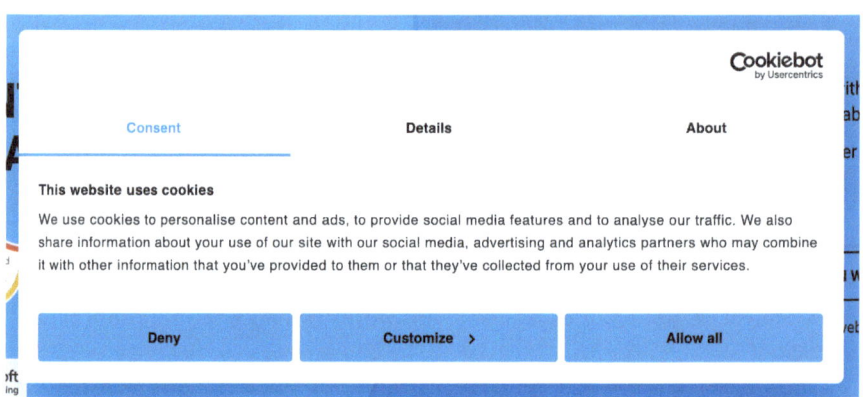

FIGURE 5.14
As a company focused on providing cookie banners as a product, Cookiebot features more user-friendly controls in their flexible "consent solution" while also ensuring companies comply with privacy laws.

The following banner from the UK independent music site the Quietus also uses Cookiebot's cookie solution. It's very large but provides the most options of those reviewed here, including the ability to proceed, deny, or customize what cookies are being used (see Figure 5.15). Like the previous banner, it presents additional information behind tabs without pressing the user to go to a separate page. Note, too, that it mentions how many partners your data is being shared with, allows you to drill down and review them, and to deselect them.

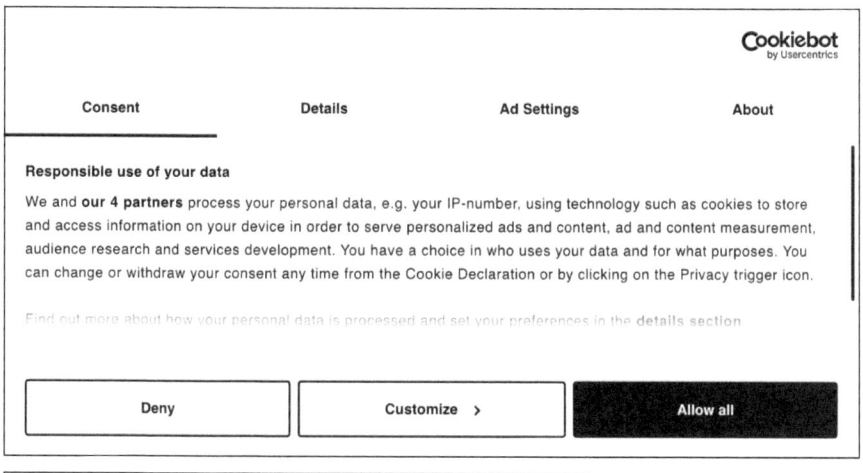

FIGURE 5.15
This cookie banner on the Quietus site informs visitors that they're sharing data with four partners for processing and allows you to click directly through from the copy or the "Customize" button to deselect them if you want.

Cookie banners make for a small, digestible pattern to review here, although they can be surprisingly complex to design, requiring hours of deliberation with stakeholders to complete. Often, as you've seen, companies implement third-party solutions to handle cookies for them. Still, keep in mind that these guidelines apply to any other design pattern you might design, hoping to obtain consent to gather users' data, too, such as onboarding or just-in-time alerts. Note, for example, in Figure 5.16, how Amazon Prime uses the same pattern for buried consent that you reviewed previously to easily enable personalized recommendations while forcing the customer to dig deeper to turn them off.

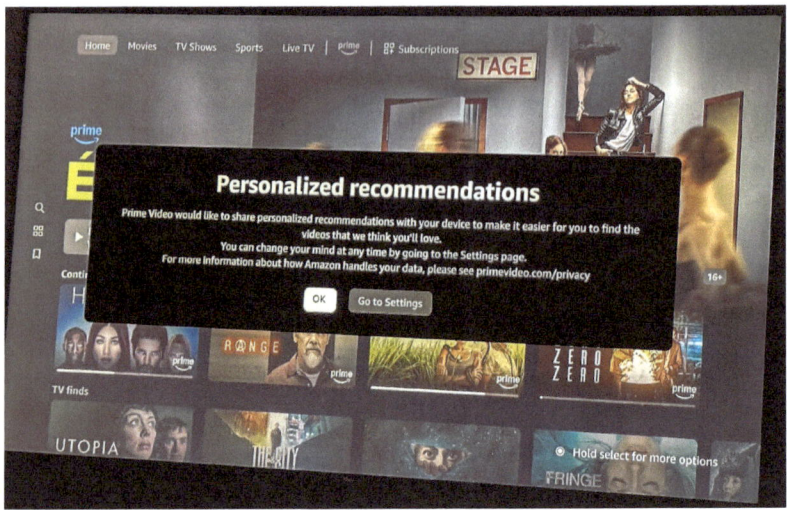

FIGURE 5.16
During onboarding to Amazon Prime's streaming video service, users can choose "OK" to proceed with personalized recommendations but must select "Go to Settings" to avoid sharing their information.

> **EXERCISE** DESIGN A PRIVACY-FRIENDLY COOKIE BANNER
>
> The last cookie banner (Figure 5.15) seems helpful, but it's also quite large. Can you think of some ways that you could convey this same information thoughtfully and effectively without taking up so much screen real estate? Can you devise a pattern that would strike an even better balance than the ones you reviewed above?

Make Sure Users Can Delete Their Data

Ensure that users can access features to completely delete their data. That also means allowing them to delete their entire account—and easily. Among many companies, Amazon has perhaps most famously caught heat from customers by making it much more onerous to delete your account than to create it. Entire articles have been written explaining how to cancel your Amazon Prime account, a distinction few other companies enjoy.

Facebook has also made it notoriously difficult to *truly* delete your account. As of this writing, it takes approximately 20 discrete steps to delete an account, the first 10 of which are simply drilling down to the point in the experience that allows you this option. (See Figure 5.17.)

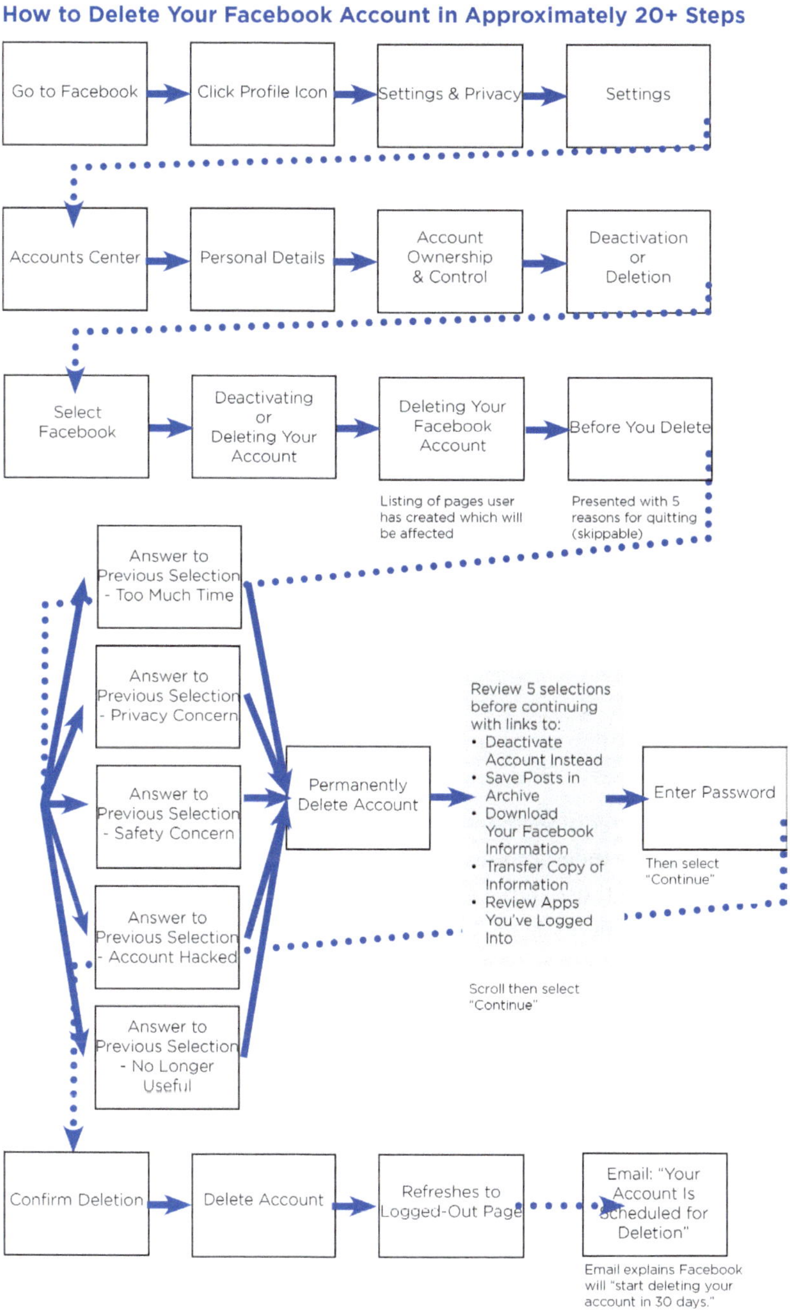

FIGURE 5.17

This diagram shows the approximately 20 steps it takes to delete your Facebook account. Steps vary according to what related Meta accounts the user has and whether they've created additional pages on the platform.

Handle Data Responsibly 79

Within the remaining steps, Facebook attempts to convince you to deactivate your account instead of deleting it, asks why you're leaving, offers solutions to your issues with the platform, tries to convince you to just deactivate (again), asks you to confirm, and so on. (See Figure 5.18.)

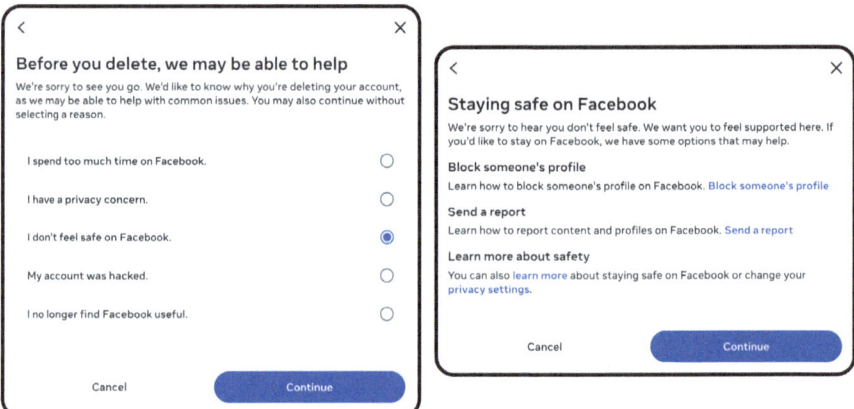

FIGURE 5.18
These Facebook screens address a user's concerns about why they're leaving the platform, adding friction to the user's attempt to delete their account.

Some new laws require companies to make it just as easy for consumers to delete their accounts as it was to create them, meaning experiences like those that Amazon and Facebook insist you step through will become illegal. Similarly, in late 2024, the FTC announced a new rule that requires companies to make it easier to end recurring subscriptions and memberships, though, as of this writing, the rule's future remains uncertain, given business and political pushback.[20] (Note, however, that ending a subscription or membership is *not* the same as deleting your account and all its related data.)

---

20 "Federal Trade Commission Announces Final 'Click-to-Cancel' Rule Making It Easier for Consumers to End Recurring Subscriptions and Memberships," Federal Trade Commission, 16 October 2024, www.ftc.gov/news-events/news /press-releases/2024/10/federal-trade-commission-announces-final-click -cancel-rule-making-it-easier-consumers-end-recurringzz

EXERCISE SIMPLIFY A BYZANTINE ACCOUNT DELETION PROCESS

Imagine that you were tasked with simplifying a large social media company's account deletion process. Assume that each profile contains a lot of personal data with images and posts potentially going back for several years. Take a moment to create a flow diagram, showing the steps you'd need to include.

Here are some key considerations:

- How specifically might you simplify the process, keeping it to 10 steps or less?
- How could you strike a balance of reducing the friction involved in deleting an account while still allowing for appropriate safeguards to prevent people from accidentally deleting their account, to allow for specific actions like saving your data first, and to prevent bad actors from deleting someone else's account without their permission?
- How might you make users feel that you support their decision while balancing that to an appropriate degree with the business's need?
- How would you reassure users that their account and all its data have, in fact, been completely deleted?

In May of 2024, the King's College London released a study of the 20 most popular period tracking apps.[21] The study found that many of the apps required users to disclose whether they had miscarried or had an abortion and failed to provide any way for users to delete data from the apps. The lead investigator of the study Dr. Ruba Abu-Salma concluded that this amounted to coercion, saying "Requiring users to disclose sensitive or potentially criminalizing information as a pre-condition to deleting data is an extremely poor privacy practice with dire safety implications. It removes any form of meaningful consent offered to users."

Companies handling such sensitive data should also provide users with the ability to delete their data at the moment the need arises. In fact, emerging privacy laws often include mandates to provide users

---

21 Dr. Ruba Abu-Salma, "Female Health Apps Misuse Highly Sensitive Data, Study Finds," King's College London, 13 May 2024, www.kcl.ac.uk/news/female-health-apps-misuse-highly-sensitive-data-study-finds

with the ability to access, update or correct, and delete or transfer their personal data. That means designers can increasingly expect to be enlisted to ensure that platforms provide functionality that allows users to complete these actions. That speaks to creating the sort of privacy-oriented tools and features you'll review in Chapter 8.

Abuse and Data Deletion

Somewhat tangentially, if you work on any sort of platform that allows users to upload other people's content (personal data, photos, etc.), you need to provide a mechanism for people to request that this material be taken down and purged from any servers. If the data referenced is PII or highly personal images (for example, revenge porn), you should also provide the means to escalate these requests quickly before greater harm can be done to the individual involved. In these cases, the individuals affected have no opportunity for consent, but the companies involved will consider the actions abuse to be countered as opposed to a data privacy issue. Nonetheless, the individuals will experience them as privacy violations.

People experiencing this level of exposure can experience unusual distress, especially when there's no clearly available means for allowing them to delete or report content, and they find themselves navigating an endless loop of emails and phone calls with an array of company employees, accompanied by a growing Kafkaesque feeling of dread that they're not really getting anywhere.

Automatic Deletion

Some companies go the extra mile and will delete your information after a set period of time—ideally, also offering you a warning or two in advance via email that they're going to delete your data or even your account.

In the following email from Khan Academy, for example, you'll learn the company has a policy of deleting accounts that haven't been used for a time, expressly to adhere to their privacy policy and "to protect our users' privacy and security." (See Figure 5.19.) They make it easy for you to maintain your account, however. All you have to do is click the prominent green button to log into the site. Logging in signals your desire to stay on board.

FIGURE 5.19
Khan Academy emails users who have not engaged with the platform for a while to say the company will delete their account soon along with their personal data as a privacy measure. Users need to log in within a couple of days to keep their accounts.

EXERCISE RESIST OVERSHARING ON A DATING APP

Assume that you're an internal employee at a company that's creating a dating app for singles to use globally. You've learned that they're collecting a lot of personal information around their users' preferences in the hope of selling that data to third parties for the purposes of creating some incredibly targeted online marketing. What are some approaches you could take to design the app's experience to ensure that you're explaining to subscribers up front what it is that they're sharing, with whom, and why? What specific design patterns might you consider?

Also, what guidance might you offer the company's senior managers and strategists who have concluded it's imperative for the company to increase their bottom line by selling their users' personal data? What limitations would you recommend placing on the use of their data?

"THE RIGHT TO BE FORGOTTEN"

A "right to be forgotten," also described as a "right to erasure" or "a right to oblivion," has been argued successfully in many countries around the world. For example, in 2014 the European Court of Justice ruled in a case against Google Spain that European citizens have a right to ask search engines to remove certain content related to their names that had become public for a period of time, especially if that information is inaccurate or no longer considered relevant but could cause them harm. That ruling set a precedent for the right to be included in the GDPR, as well. France had already put *le droit à l'oubli* into effect in October 2010. With precedents going back to the 1990s, however, Argentina, is currently considered to have the strongest right to be forgotten laws on the books—to the degree that some critics argue they could be leveraged to remove content from view that the public *should* have the right to see.

Despite majority support for it, this right has not gained traction as quickly in the United States, where some argue it conflicts with the right to free speech and a "right to know." Google, specifically, has a history of resisting a right to be forgotten. Writing on the topic for his personal blog in 2011, the company's global privacy counsel Peter Fleischer concluded that his conviction remained unchanged: "History should be remembered, not forgotten, even if it's painful. Culture is memory."[22] People often assume that requests to have their mentions deleted from Google results come from violent criminals and embarrassed celebrities, but in 2015 *The Guardian* conducted research that concluded the truth is very different and much less "sensational":[23]

---

22 Peter Fleischer, "Foggy Thinking About the Right to Oblivion," *Peter Fleischer: Privacy...?* (blog), 9 March 2001, https://peterfleischer.blogspot.com/2011/03/foggy-thinking-about-right-to-oblivion.html

23 Sylvia Tippmann and Julia Powles, "Google Accidentally Reveals Data on 'Right to Be Forgotten' Requests," *The Guardian*, 14 July 2015, www.theguardian.com/technology/2015/jul/14/google-accidentally-reveals-right-to-be-forgotten-requests

Less than 5% of nearly 220,000 individual requests made to Google to selectively remove links to online information concern criminals, politicians and high-profile public figures, the Guardian has learned, with more than 95% of requests coming from everyday members of the public.

Similarly, in her book *Algorithms of Oppression*, Safiya Umoja Noble concludes, "Google's control and circumvention of privacy and the right to be forgotten intensifies damage to vulnerable populations."[24]

Remember that those seeking a right to be forgotten often pursue it for understandable reasons. Women and girls, in particular, may find their names and images have been linked to pornographic material illegally or without their permission. Some people may be victims of crime or witnesses to it. Some may be attempting to protect children from content that highlights their personal information. And some individuals may have committed crimes in their past, but have served their time, taken their lives in new directions, but may still find it difficult to secure employment if search results reveal their earlier convictions. Companies which dismiss any sort of right to be forgotten have not carefully considered the impact of their practice upon already at-risk populations—nor people with specific needs in the broader population.

Imagine, too, an ultimate nuclear option for online privacy: To have any mention of oneself whatsoever purged from the internet altogether. We live in an age now where that's practically impossible. That's another reason it's imperative for people to have some measure of control over their own data, as well as other information about them. The internet, as it's said, is forever.

---

24 Safiya Umoja Noble, *Algorithms of Oppression* (New York: New York University Press, 2018), 130.

The Takeaway

By applying this level of care to procuring and securing people's data, you build users' trust, a significant consideration for maintaining a business's brand integrity. If, as a designer, you take care to examine what data points are truly needed, if you take pains to communicate the what, why, and who of sharing, and if you ensure clear and prominent moments arise for people to signal their consent, you've now positioned yourself nicely to create an experience that should prove to be safer and more appealing to your users. Don't forget, however, to include those features that allow users to withdraw their consent and to delete their data or even their whole account.

Now, proceed with confidence!

CHAPTER 6

Avoid Deceptive Patterns

Defining Deceptive Patterns	88
Deceptive Patterns Affecting Privacy	89
Approaches for Avoiding Deceptive Patterns	109
The Takeaway	118

When you encounter deceptive patterns online, you can assume skullduggery is afoot. In *Ruined by Design*, design director Mike Monteiro suggests that deceptive or "Dark patterns are the canaries in the coal mine of unethical design. A company who's willing to keep a customer hostage is willing to do worse."[1]

Acquainting yourself with these particularly harmful or even hostile patterns plays a significant role in designing for privacy because, as you grow familiar with them, you'll soon be able to spot them in their various forms and know how to avoid them, advise against them, and suggest better alternatives to your company or clients.

After considering a definition for "deceptive patterns," you'll review some specific patterns that are often used to undermine users' privacy:

- Bait and switch
- Bad defaults
- Privacy Zuckering
- Forced actions
- Hidden options
- Cookie consent manipulation
- Interference

Then you'll consider some approaches for avoiding these harmful patterns.

Defining Deceptive Patterns

The UK-based UX designer and expert witness Harry Brignull coined the term *dark pattern* in 2010 to describe patterns designed to intentionally deceive, coerce or manipulate. Harry and his team later changed the term to *deceptive pattern* to "reflect a commitment to avoiding language that might inadvertently carry negative associations or reinforce harmful stereotypes."[2] More broadly now, Brignull speaks to manipulative or deceptive patterns, since "deceptive" can be a very narrowly defined term.

---

1 Mike Monteiro, *Ruined by Design* (Mule Design, 2019).
2 Harry Brignull et al., "Deceptive Patterns—User Interfaces Designed to Trick You," deceptive.design, 25 April 2023, **deceptive.design**

> **NOTE** **DECEPTIVE PATTERN**
>
> This book uses the term *deceptive pattern*, too, but since you'll see designers and scholars reference both terms elsewhere, consider this confirmation that both terms refer to the same type of patterns.

Earlier you learned that California has banned specific deceptive patterns. Recently, the EU has sought to do the same. You can assume a general trend toward banning these harmful patterns. Companies such as Google have already been fined for implementing deceptive patterns that undermine users' privacy.

In the context of privacy by design, think of deceptive patterns as those patterns designed intentionally to trick people into performing actions they didn't intend to do, especially as it relates to sharing their personal information. Interacting with these patterns can significantly undermine users' privacy. Indeed, these patterns are often designed to do precisely that.

Deceptive Patterns Affecting Privacy

A number of different taxonomies have been developed to describe and categorize deceptive patterns. The following deceptive patterns have been identified from across these frameworks. These examples should *not* be considered exhaustive, but they do showcase several patterns most likely to disrupt people's privacy. Take a moment to study them and don't allow them to creep into your designs.

Bait and Switch

Bait-and-switch patterns do exactly what you'd expect: promise you one thing but deliver another. In some cases, you, the user, may be presented with a completely different product than you were expecting or via "sneaking," a company may add something to your order you didn't select. In others, you may take the bait and find that you have to offer more information than you were originally told.

The following example from the Peak Design website showcases an egregious example of one sort of bait-and-switch pattern that tricks potential customers into providing more of their personal information than they initially thought. It opens by promising you $10 off a $75 purchase and offers a single welcoming field for your email address.

Step 1: The Bait: The experience presents visitors with a pop-up that appears to offer them an amount off their purchase in exchange for their email address. (See Figure 6.1.)

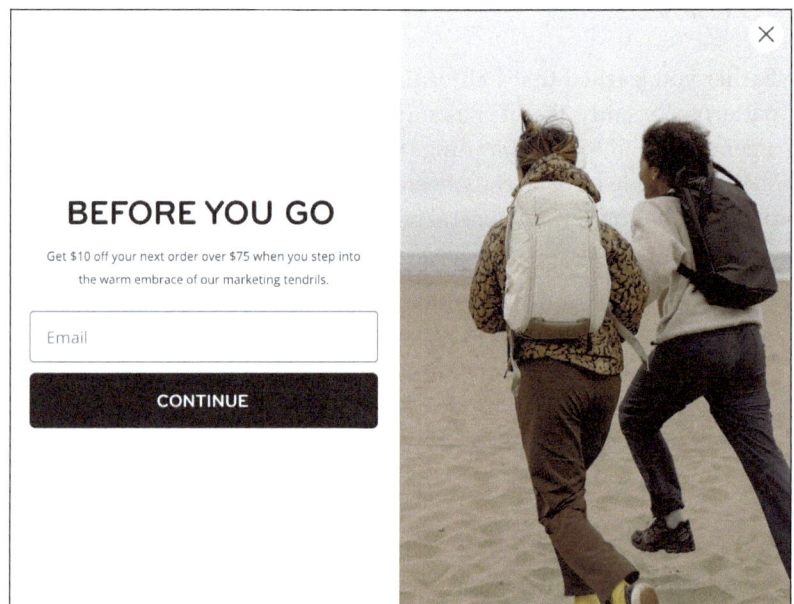

FIGURE 6.1
The first pop-up on Peak Design's site appears to request just your email address in exchange for an offer of $10 off.

Step 2: The Switch: The site then presents visitors with a second and unexpected state of the pop-up, asking them to supply their mobile number *as well* to secure the offer. (See Figure 6.2.)

Now, if the visitor decides not to proceed by entering their phone number, the company has already secured their email address, but the visitor doesn't get the discount. If the visitor does proceed, the company secures their personal phone number as well. Now the visitor has surrendered two key pieces of their personal data without even having made a purchase yet. Note, too, that the user's mobile number is requested after the email address. You'll usually find that order reflected in instances of this pattern because, clearly, a phone number is a data point that feels notably more personal than an email address. Visitors are slightly more invested after they've supplied their email address, so they are slightly more likely to add their phone number as well.

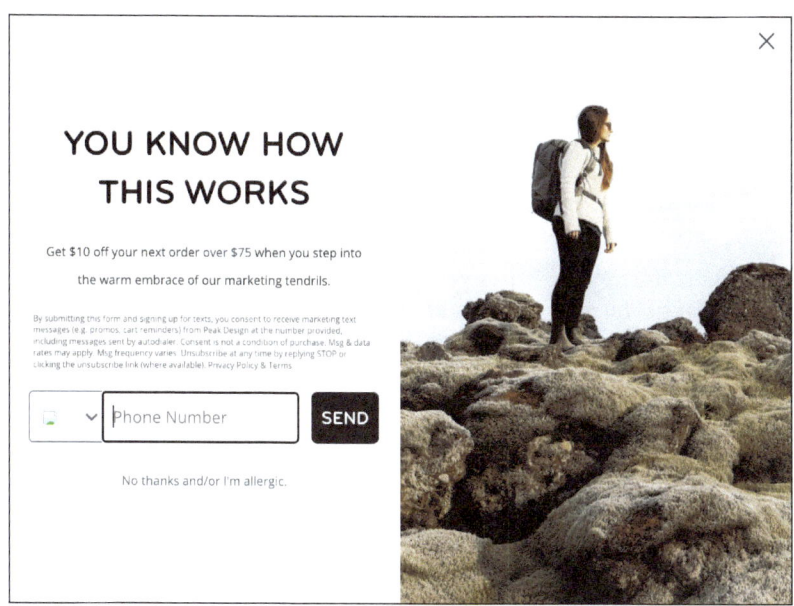

FIGURE 6.2
Surprisingly, the second screen reveals that you have to submit your phone number as well to secure the discount.

No visitor would reasonably expect that they would still need to surrender their mobile number after entering their email address in this scenario. In retrospect, you can see the copy in the first pop-up was carefully crafted to avoid telling an outright lie. Still, you can only tell that after you've continued by submitting your email address. The banner's jokey, informal copy promises that you'll get $10 off "when you step into the warm, loving embrace of our marketing tendrils." The tone is funny and a little self-deprecating. But you would likely assume that if you simply enter your email address, the 10 bucks will be bestowed upon you. The pattern was calibrated to secure more of your personal information than you imagined. It's intentionally deceptive.

Interestingly enough, patterns for handling a discount honestly prove much simpler to design and develop. Here's how Langly, a competitor to Peak Design handles a similar offer in Figure 6.3.

AVOID DECEPTIVE PATTERNS

FIGURE 6.3
Langly uses simple, straightforward copy within their offer modal: "Get Discount Now." True to their word, visitors are presented with a discount code that is also sent to the provided email address. No phone number is required. (Guess which site the author bought a camera strap from?)

Of course, the user who secures Langly's discount code can also expect to receive future marketing emails from the company, but, at least, the exchange is fairly transparent. Hopefully, those emails will include a simple way to "unsubscribe," as well.

Bad Defaults

You've likely encountered innumerable preselected checkboxes within experiences you've visited online. Those serve as one of the most common deceptive patterns users encounter, part of a larger pattern known as *bad defaults*. You may also see this practice referred to as *preselection* (see Figure 6.4). The pattern could be used for anything from ensuring that you've unwittingly subscribed to a newsletter, tricked you into an upsell, or from a privacy perspective, shared your entire collection of contacts with a company only for them to send an email to all of your friends and associates en masse. Bad defaults often employ radio buttons, too.

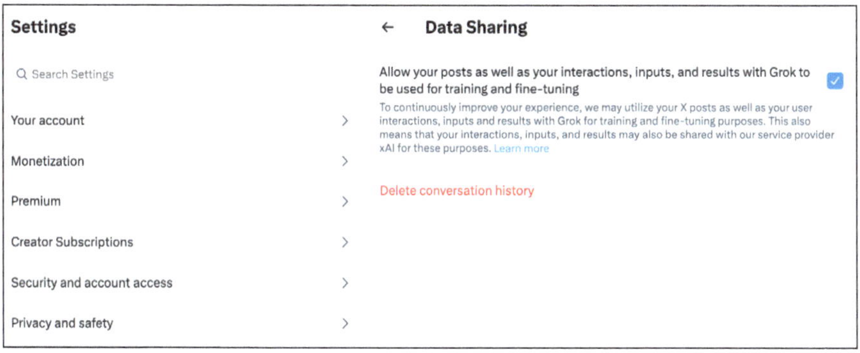

Settings		← Data Sharing	
Q Search Settings		Allow your posts as well as your interactions, inputs, and results with Grok to be used for training and fine-tuning	✓
Your account	›	To continuously improve your experience, we may utilize your X posts as well as your user interactions, inputs and results with Grok for training and fine-tuning purposes. This also means that your interactions, inputs, and results may also be shared with our service provider xAI for these purposes. Learn more	
Monetization	›		
Premium	›	Delete conversation history	
Creator Subscriptions	›		
Security and account access	›		
Privacy and safety	›		

FIGURE 6.4

On the social media app X, a feature allows the platform to share all of a user's posts, interactions, inputs, and results with Grok, the platform's AI "for training and fine-tuning purposes." The checkbox for this feature is preselected and buried deep in X's settings.

VENMO: PUBLIC BY DEFAULT

Bad defaults can prove incredibly damaging when the practice leads to making users' activity itself on a platform *public by default*. This usually involves a setting turned to "public" by default somewhere in the experience, leading them to sharing likes or preference or activity far more broadly than they ever realized. Often, users can fix the issue by changing a setting from "public" to "private" when it should have been private by default. Leading with "public," however, is opposite of designing with privacy as a default. Some experiences may not even offer the ability to turn off public activity at all.

To this day, when you make a payment on Venmo, your posts there default to public, so you automatically share your payments with…everyone. (See Figure 6.5.)

continues

VENMO: PUBLIC BY DEFAULT (continued)

WHO CAN SEE MY TRANSACTIONS?	When you pay or get paid on Venmo, you decide who can see it. You can select the settings, including Public (visible to everyone on the Internet), Friends (visible to sender, recipient, and their Venmo friends), or Private (visible to sender and recipient only). Payment amounts are always secret.
	Your initial default privacy setting is Public, so everyone on the Internet can see your payments. You can change it for each transaction individually, or for all in Settings. If you and your friend use different privacy settings, we use the more restrictive one when you're paying each other.

FIGURE 6.5
On a screen entitled "Helpful Information," Venmo explains, "Your initial default privacy setting is Public, so everyone on the Internet can see your payments."

Around 2015, two developers Mike Lacher and Chris Baker created a website called Vicemo that scraped payments listed with words associated with "drugs," "sex," and "alcohol" and posted them online for anyone to see (see Figure 6.6). Of course, many of these people may just have been joking with their friends when they included "mushrooms" or "smack" in the details field of their Venmo payment.

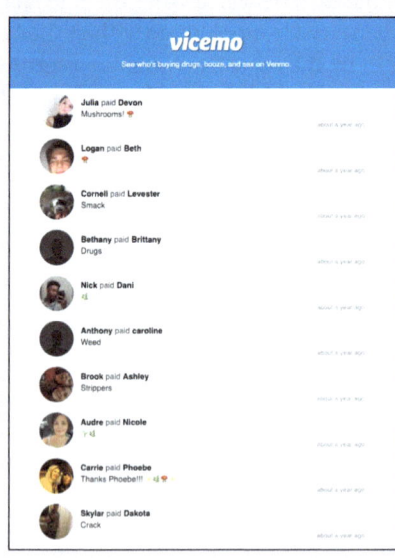

FIGURE 6.6
Vicemo, the website that presented scraped Venmo payments that referenced sex, drugs, and alcohol.

In 2018 Joel Guerra created @venmodrugs or "Who's buying drugs on Venmo?," a Twitter account that provided a similar view into Venmo posts. Both of these efforts were initiated by self-proclaimed "white hat" hackers who wanted to demonstrate how this public information looked when presented in another context. Both projects presented data that anyone could see in the social feed of the app. The projects merely highlighted that fact.

Venmo has long received criticism for this pattern, but the experience essentially remains the same, although Venmo now allows users to change the setting for each transaction to private or to change all their transactions to private in their settings. (See Figure 6.7.)

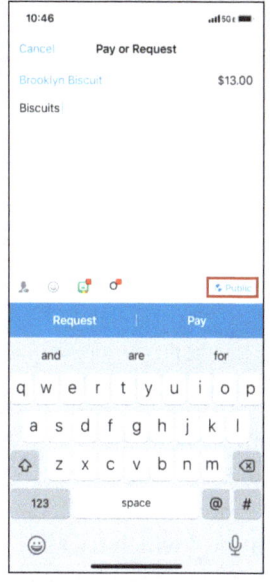
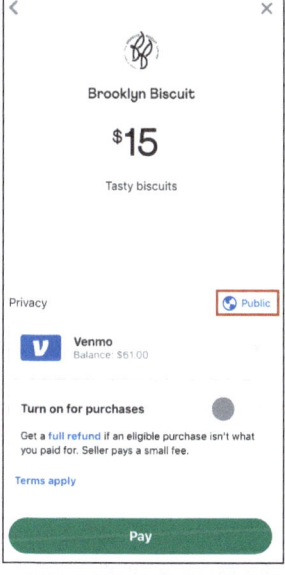

FIGURE 6.7

Two variations of the Venmo payment screen over time, both showing how payments default to "Public," unless the customer changes the setting with each transaction or digs into settings to change the public default to private permanently.

Nonetheless, that setting remains public by default to this day. It should have always been private. In fact, in recent months, two politicians in the U.S., Pete Hegseth and Matt Gaetz, both had potentially damaging information made public based upon the history of their Venmo transactions.

Privacy Zuckering

Tim Jones landed upon the term *Privacy Zuckering* in 2010 to label the practice of burying what's really happening with users' personal data in a privacy policy or terms of use statement, so that users can't really be sure what's happening with their information. Unsurprisingly, Jones had Meta CEO Mark Zuckerberg and Facebook in mind when he coined the term.

Upon closer examination, *Zuckering* often uses a few specific deceptive patterns to accomplish its directives, particularly the use of opaque or deceptive language. Additionally, platforms may offer users tools to adjust their privacy settings that prove incredibly complicated or detailed, so that they become exasperated and give up on maintaining control of their own data.

Similarly, Zuckering may also incorporate "bundled consent," which tricks people into sharing more than they intended by combining different types of data sharing within a single explanation or selection.

The nuances of data sharing are often buried in privacy policies, too. Many people, for example, are not aware that Meta owns WhatsApp. If you dig into the privacy policy for the encrypted messaging app, however, you'll find that "WhatsApp receives information from, and shares information … with, the other Meta Companies." (See Figure 6.8.)

Forced Actions

A forced option pattern pushes users down a path without the indication that alternate paths may be available to them. Sometimes, alternate paths may exist, but they're disguised or not mentioned. A typical example of this pattern is "forced registration," which tricks users into creating an account or profile when they didn't need to. As researcher Dr. Colin M. Gray and their team at Indiana University's UX Pedagogy and Practice Lab (UXP2) explain in their study, "The Dark (Patterns) Side of UX Design," forced options can take on different forms. They highlight "social pyramids," which require you to recruit other people to continue, "gamification," which forces you to pay to level up within a game, and the aforementioned "Privacy Zuckering," which forces users to share more personal data than they intend to.[3]

---

3 Colin M. Gray et al., "The Dark (Patterns) Side of UX Design" (presentation, CHI 2018, Montréal, QC, Canada, April 21–26, 2018).

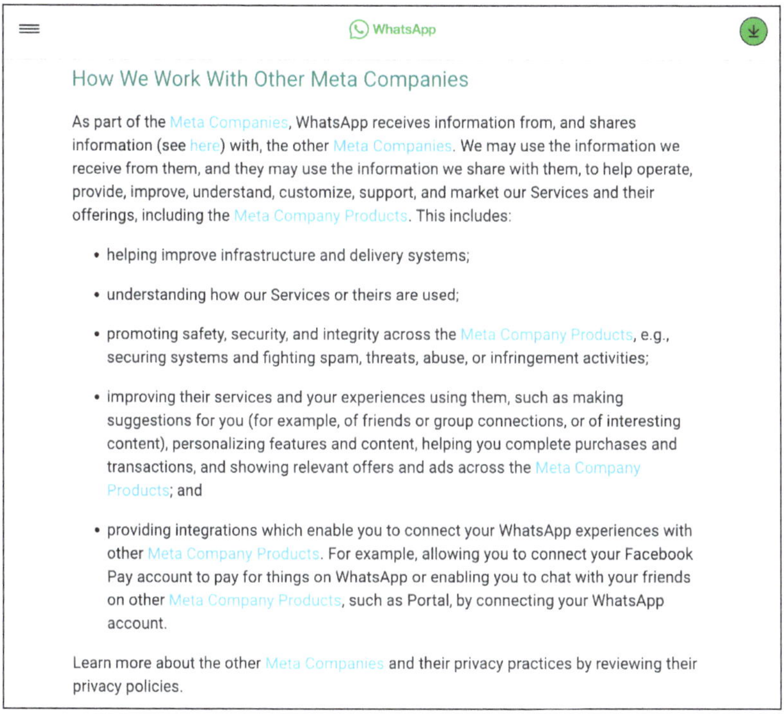

FIGURE 6.8
Some users may not even know that WhatsApp is owned by Meta. Either way, many users might not guess that elements of their data are being shared across a greater array of Meta properties.

The luxury resale site RealReal requires users to submit an email to sign up before they can view the details of any of their products (see Figure 6.9). Alternatively, users can sign up with an Apple, Facebook, or Google account. If a user proceeds with Facebook, that likely means they will be sharing Facebook data such as their friends' list, likes, and other preferences with RealReal. Additionally, Facebook will likely be able to track their visits with that business from that point on.

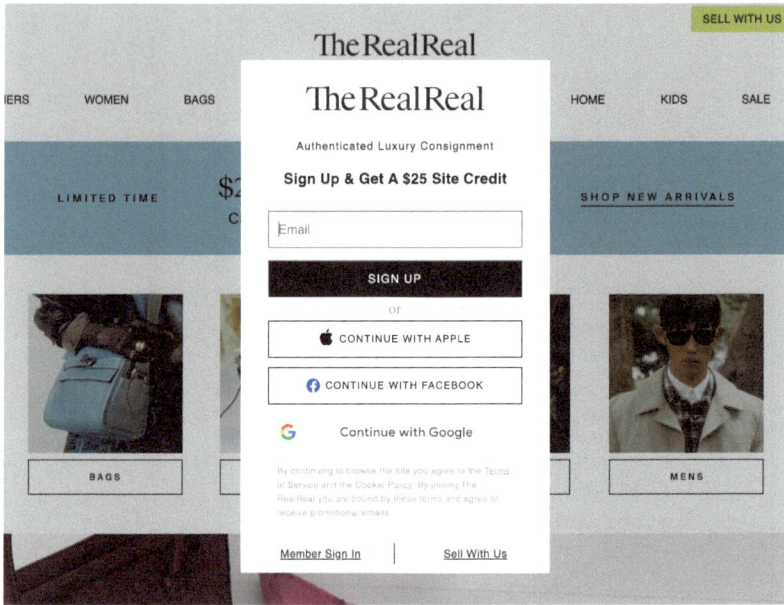

FIGURE 6.9
The RealReal site features luxury consignment products, but it requires potential customers to create an account before seeing product details.

Upon its launch, for example, the photo-sharing app Lapse angered many users, when they discovered they were required to invite five friends to the app before they could use it (see Figure 6.10).[4] That's an example of "growth hacking," which is a practice where companies use various, sometimes shady, devices to grow their user bases as quickly as possible. It also provides an illustration of how these deceptive patterns are often used in different combinations to accomplish a particular goal.

---

4 Sarah Perez, "Photo-Sharing App Lapse Hits Top of the App Store by Forcing You to Invite Your Friends, TechCrunch," 26 September 2023, https://techcrunch.com/2023/09/26/photo-sharing-app-lapse-hits-top-of-the-app-store-by-forcing-you-to-invite-your-friends

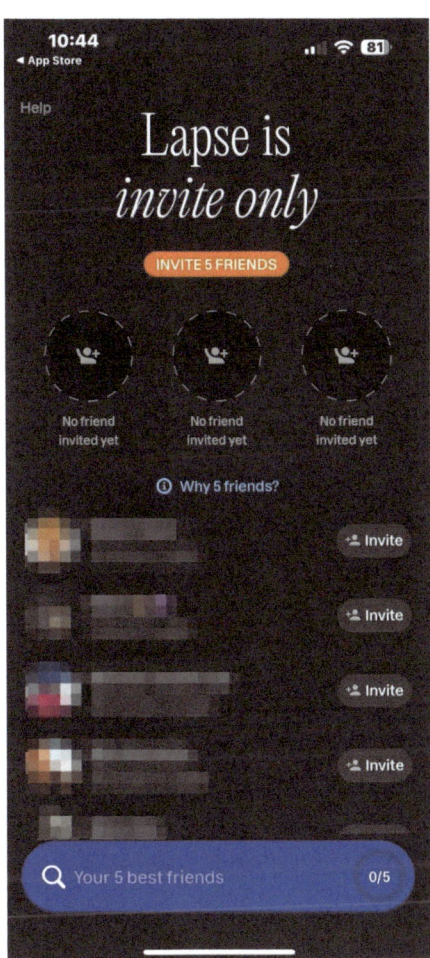

FIGURE 6.10
This screenshot from Lapse featured in a TechCrunch article about the app demonstrates how new users were asked to invite five friends before proceeding.

LINKEDIN AND GROWTH HACKING

"I will never forget how I felt when LinkedIn tricked me into inviting everyone on my Gmail contact list to join.... It was terrible! My contacts included old teachers, customer service representatives, business contacts, extended family, and many others who did not appreciate the spam it sent out on my behalf. I felt embarrassed and betrayed."—Jonathan Shariat in *Tragic Design*.[5]

Several years ago, LinkedIn presented potential members with a doozy of a deceptive pattern when they were signing up for the networking platform. Essentially, LinkedIn suggested you import your contacts so they could "Invite them to LinkedIn so they can connect with you." The modal showed all of your contacts as preselected, and the most prominent CTA button read "Add to Network." If you didn't read the copy very carefully, you might not realize that LinkedIn was about to email everyone in your address book. (Experiences often incorporate features that look almost identical to this, but only to help you find people you know who have *already* joined the platform.) Imagine how mortifying it would have been to discover that you had just spammed everyone with an invite to LinkedIn in your contacts list—potentially hundreds of people?

The former Googler and senior product manager Dan Schlosser wrote an exhaustive piece detailing this pattern after he discovered he'd fallen prey to the issue himself, saying he felt LinkedIn had tricked him into sharing the entire content of his address book.[6] (See Figure 6.11.)

Schlosser demonstrated in his piece that it wasn't just a single pattern either: LinkedIn created a thicket of deceptive patterns presented on multiple screens, all designed to trick you into surrendering your personal contacts, even if you had skipped sharing them earlier in the sign-up process.

LinkedIn used, not just preselected checkboxes, but multiple forms of deceptive patterns in this exercise. For example, you could choose only "skip" but never "deny" when sharing your contacts. That represented a subtle but important distinction because it allowed LinkedIn to bug you repeatedly

---

5   Jonathan Shariat and Cynthia Savard Saucier, *Tragic Design* (O'Reilly Media, 2017).
6   Dan Schlosser, "LinkedIn Dark Patterns," *Medium*, 5 June 2015, https://medium.com/@danrschlosser/linkedin-dark-patterns-3ae726fe1462

afterward to share your contacts anyway. Also, if you reviewed all the screens involved, you would find that LinkedIn used slippery or straight-up deceptive language as well to trick people into supplying their contacts.

This feature capitalized on bad defaults and forced action patterns. This specific type of deceptive pattern is referred to as *address book leeching*,[7] and it's often used for "growth hacking" or "forced invites," both of which platforms deploy in the hopes of growing their user bases as quickly as possible.

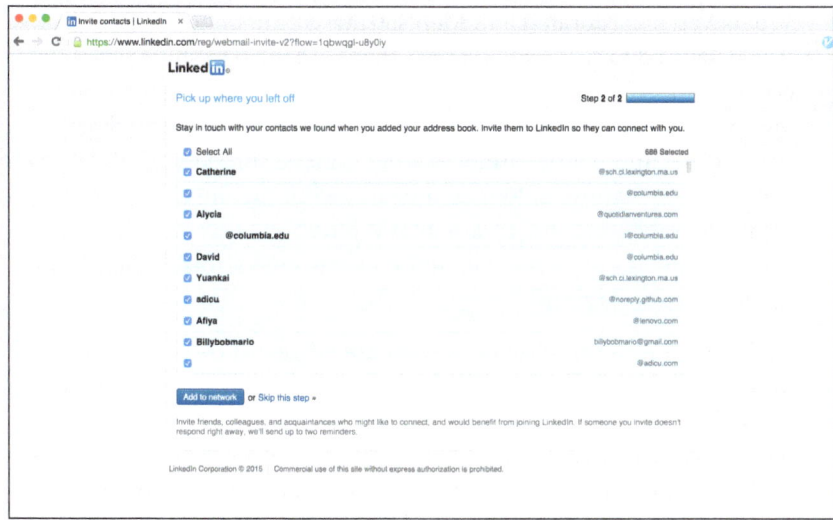

FIGURE 6.11
Note how the feature was preset to invite all 688 of Dan's contacts to LinkedIn if he had simply selected "Add to network" instead of the deprioritized "Skip this step" button.

---

7   Christoph Bösch et al., "Tales from the Dark Side: Privacy Dark Strategies and Privacy Dark Patterns," *Proceedings on Privacy Enhancing Technologies*, July 2016: 237–254.

Hidden Options

Forced action patterns often use "hidden options," a deceptive pattern itself that buries settings that would better maintain your privacy—provided you were able to find them. The point of this pattern is to provide these possibly required options in places that are not intuitive to find or are planted so deeply that users aren't likely to happen across them. The European Data Protection Board refers to patterns like this as a *Privacy Maze* and references GDPR provisions it violates, such as transparency, fairness, and informed consent.[8]

Privacy Zuckering often makes use of hidden options. Previously, WhatsApp incorporated a hidden setting within onboarding that was defaulted to "on" and allowed the WhatsApp to share users' account information with Facebook "to improve my Facebook ads and products experiences." (See Figure 6.12.) It did also say that users' chats and phone numbers would not be shared with Facebook. This sharing feature no longer appears in the current app. However, it's unclear whether this means that WhatsApp is *not* still sharing certain elements of users' data to fine-tune advertising presented on Facebook or Instagram.

In the previous chapter, you learned how difficult Facebook users find it is to cancel their accounts, an issue that showcases another example of deeply hidden options. Similarly, Amazon was shown to have ignored feedback that would have made canceling Prime easier. Instead, the company intentionally created a confusing, convoluted cancellation labyrinth that they referred to internally as the "Iliad Flow," after Homer's epic 15,000 plus line poem. The FTC ruled that Amazon "required consumers intending to cancel to navigate a four-page, six-click, fifteen-option cancellation process."[9] If you wanted to sign up for Prime, however? That took just one or two clicks. In anticipation of the lawsuit, Amazon did shorten and simplify the cancellation process. Nonetheless, they were still forced to pay a penalty. Currently, it takes about five clicks to cancel Amazon Prime.

---

[8] "Guidelines 03/2022 on Deceptive Design Patterns in Social Media Platform Interfaces: How to Recognise and Avoid Them," European Data Protection Board, 24 February 2023, www.edpb.europa.eu/our-work-tools/our-documents/guidelines/guidelines-032022-deceptive-design-patterns-social-media_en

[9] Federal Trade Commission v. Amazon.com, Inc., Case 2:23-cv-00932 (USDC WD WA 2023).

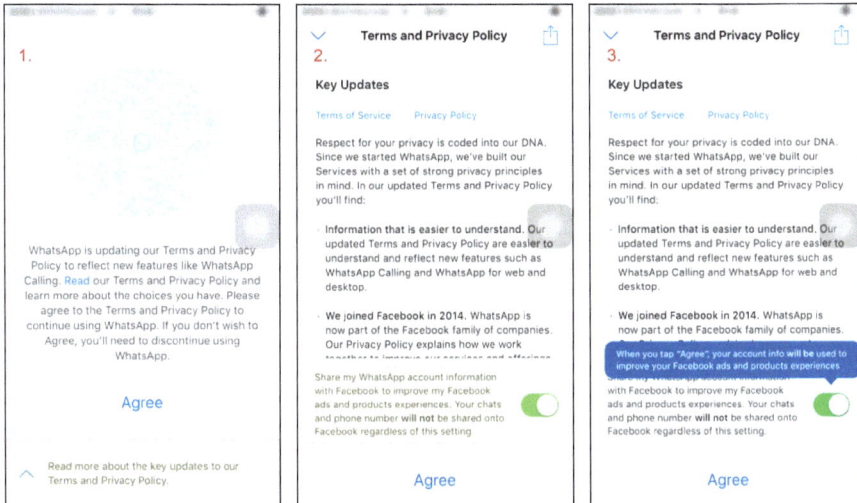

FIGURE 6.12
Screenshots from 2017 show a hidden setting allowing WhatsApp to share users' data for advertising by default. From "Privacy Zuckering: Deceiving Your Privacy by Design" by Mohit on Medium.

GOOGLE'S SYMPHONY OF DECEPTIVE PATTERNS

In 2021, while providing expert witness against Google's handling of users' location data, Indiana University's Dr. Gray argued that users would be unlikely to find important privacy settings, including those buried in Google's Privacy and Terms policy, if they didn't already know they existed. Additional research suggested that Google tracked users, regardless of their settings, and utilized the location of users who had opted into location sharing to pinpoint the location of those who had not.[10] Offering her expert witness, Stanford privacy and data policy fellow expert Jennifer King concluded, "Further, despite the various settings, there is no 'opt-out' and there is nothing users can do to prevent Google from doing this."[11]

continues

---

10 Jerod Macdonald-Evoy, "Google Court Docs Show That Users Who Opt Out of Tracking Are Still Monitored," 9 September 2022, **www.rawstory.com/google-tracking**

11 Ibid., 103

GOOGLE'S SYMPHONY OF DECEPTIVE PATTERNS (continued)

Similarly, Gray argued that Google hid features from users that allowed them to prevent the company from saving their web and app activity against their account.[12] (See Figure 6.13.)

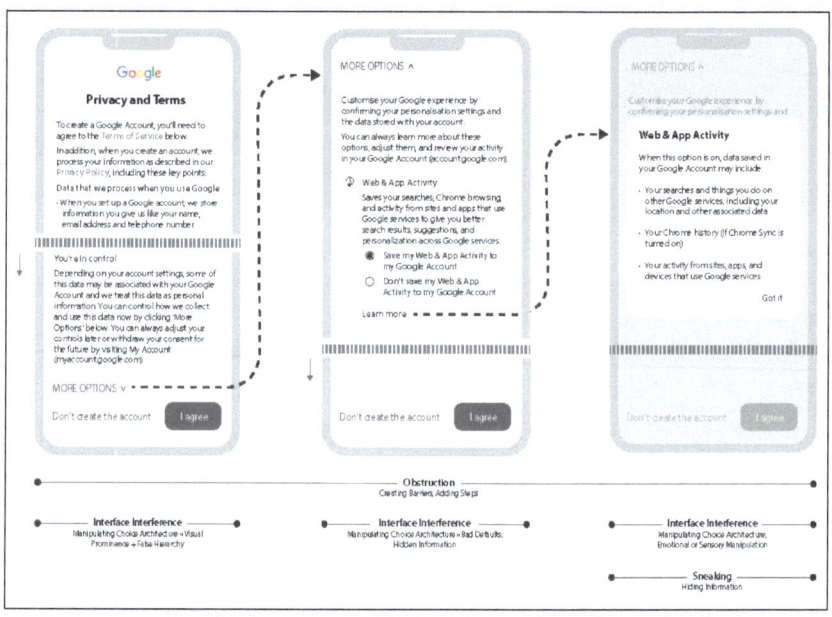

FIGURE 6.13

Illustration from Gray, Gairola, and Mildner's "Mapping the Temporal Dynamics of Dark Patterns: A Multi-Scale Approach to Analyze Dark Patterns" demonstrating how Google allegedly hid features that allowed users to deny the company the ability to save their web and app activity against their Google account.

---

12 C. M. Gray, R. Gairola, and T. Mildner, "Mapping the Temporal Dynamics of Dark Patterns: A Multi-Scale Approach to Analyze Dark Patterns" in *Dark Patterns & Manipulative Design: Conceptualising and Systematising a Key Contemporary Phenomenon from a Legal Perspective and Beyond*, ed. R. Gellert, H. Schraffenberger, and C. Santos (Edward Elgar Publishing, Fall 2025).

Gray argued that Google deployed three different high-level deceptive patterns to ensure that they could track users' real-world locations: "obstruction," "interface interference," and "sneaking." (See Figure 6.14.) Hidden options could be considered a specific form of sneaking. And interface interference could be used to hide such options, as well. All those deceptive patterns worked together to create an opaque privacy experience for users.

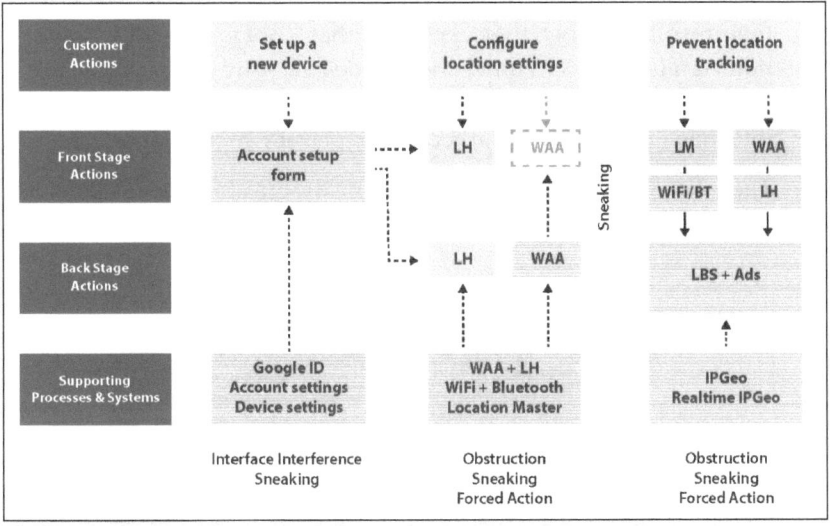

FIGURE 6.14
Providing this diagram to illustrate their point, Gray, Gairola, and Mildner argue that Google created a thicket of deceptive patterns across three different customer actions to ensure that they could still track users' real-world locations.

A counterargument to this criticism of Google's privacy experience would be that the company is required to present a tremendous amount of content to users, often including legal language, and that the struggle to meet the needs of the business, privacy laws, and consumers results in a complex experience that, however confusing, isn't intended to be a deceptive pattern. Nonetheless, extensive courtroom testimony and reporting appear to suggest that Google has historically worked hard to pinpoint users' locations, even when they've explicitly requested not to share that information.

Cookie Consent Manipulation

As you've seen, cookie banners are intended to obtain consent, but, ironically, they are often carefully crafted to trick people into consenting. Usually, they just make accepting all the cookies the simplest option, while making it more complicated to deny cookies or failing to display a way to deny them at all.

Four Norwegian researchers looked at 300 "data collection consent notices" or cookie banners implemented by news organizations and discovered that many of them incorporated deceptive patterns. In their resulting paper, they explained that "cookie consent manipulation" is a form of "circumvention by design"—by design because these cookie banners are crafted intentionally to get around the intended effects of the GDPR or other privacy regulations.[13]

The following cookie banner for a type design studio showcases this sort of manipulation (see Figure 6.15).

FIGURE 6.15
This cookie banner was featured in an article on creative examples of cookie consent experiences. Nonetheless, it incorporates the typical elements of cookie consent manipulation, which ensures that it's easier to proceed with cookies and more onerous to continue without them.

By now, you recognize the issues with this pattern already: Primarily, there's no immediate way to reject cookies, and the color and size of the "Agree & Close" button emphasize that selection.

---

13 Than Htut Soe et al., "Circumvention by Design—Dark Patterns in Cookie Consent for Online News Outlets" (paper, 11th Nordic Conference on Human-Computer Interaction, 26 October 2020).

Interference

When experiences distract you by cluttering your screen or disguising options a company would prefer you didn't take, they're implementing a deceptive pattern—better known as *visual interference* or, more broadly, *interface interference.* To distinguish between these two types of deceptive patterns, think of visual interference as using visual cues such as font or button size and color to draw your attention to desired selections. Other options may be disguised, perhaps via a smaller link with a smaller font or with low contrast colors, as opposed to an equal-sized button or CTA. Interface interference may entail deprioritizing the placement of selections or even hiding selections that the business would prefer you didn't make. Both these forms of interference seek to disguise or even hide options that should be equally weighted but are instead imbalanced, displayed via a misleading hierarchy.

Many cookie banners incorporate visual interference to nudge you toward accepting as many cookies as possible. The ways these deceptive interference patterns can be used are myriad, however. In the example in Figure 6.16, the Arc browser suggests you import your tabs, passwords, history, and bookmarks from any other browsers you may have installed.

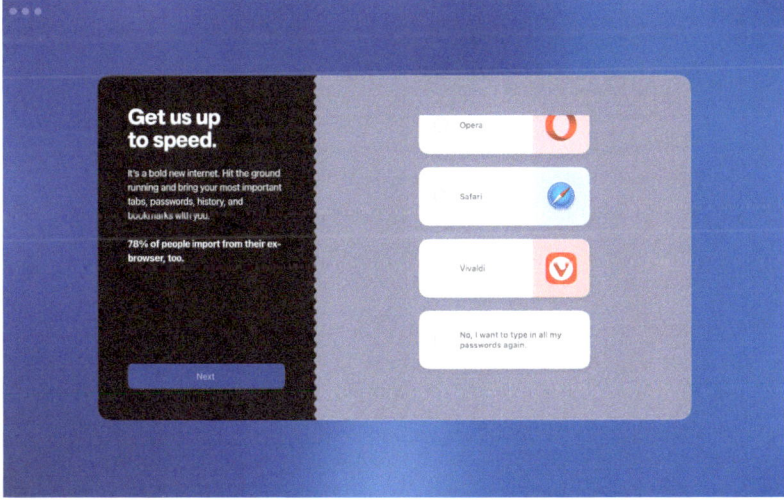

FIGURE 6.16
The Arc Browser utilizes both interface interference and confirmshaming to entice new users into sharing data associated with their other browsers.

If you don't mind sharing all your information with another company, that *could* be considered a helpful feature but note that you can't skip this screen without scrolling though all of the suggested browsers and then selecting "No, I want to type in all my passwords again." That wording, by the way, qualifies as another form of deceptive pattern you've likely encountered many times: *confirmshaming*. You can see how this intentionally complicated pattern interferes with the expected interface.

You can likely find examples of visual and interface interference in most of the patterns you've already reviewed, too. These examples betray the lengths that companies may go to in order to complicate an experience that could easily have been designed to present key options more simply and transparently.

AN ONTOLOGY OF DECEPTIVE PATTERNS

The researchers Drs. Colin M. Gray, Nataliia Bielova, Cristiana Santos, and Thomas Mildner developed a detailed "ontology of dark patterns," which they published at CHI 2024. This helpful framework for understanding deceptive patterns ladders very specific patterns up to five primary strategies or high-level patterns:

1. **Obstruction:** Intentionally making an experience more difficult than it needs to be.
2. **Sneaking:** Hiding specific information or actions from users.
3. **Interface interference:** Designing the user interface to promote certain actions over others.
4. **Forced action:** Making a user commit certain actions or provide information to access features.
5. **Social engineering:** Strategies such as false scarcity or confirmshaming that place emotional pressure on users to convince them to perform specific actions.

The patterns you're reviewing in this book fall across different levels of this framework and were selected to highlight their specific impact on privacy. Take a look at the whole framework for a more detailed understanding of how deceptive patterns work, often in concert with one another, especially since so many of these strategies can be used in ways that prove harmful to privacy.

Framework Source: ontology.darkpatternsresearchandimpact.com

That's Not All…

The patterns you've reviewed aren't the only deceptive patterns that could affect people's privacy. The are just some examples of those that affect privacy the most.

The use of deceptive language would certainly qualify as a deceptive pattern, but you'll review that privacy-undermining tactic as part of the larger concern of the importance of language in maintaining privacy in Chapter 7, "Use Language with Care." Similarly, in their paper "Tales from the Dark Side: Privacy Dark Strategies and Privacy Dark Patterns," Christoph Bösch et al. reference how "hidden legalese stipulations" use privacy-related copy to intentionally deceive.[14]

Bösch et al. also include "immortal accounts" and "shadow user profiles" in their listing of deceptive patterns that affect privacy. When designing for user experiences, consider those practices that designers can address by advising against them if necessary, but also being transparent about how data is used (Chapter 5, "Handle Data Responsibly") and by creating tools that ensure that people have control over their data (Chapter 8, "Provide Tools for Enabling Privacy").

Approaches for Avoiding Deceptive Patterns

If deceptive patterns can be deployed to undermine users' privacy, designers can reference other patterns to help reinforce it. You might look to anti-patterns and bright patterns to better understand what performs poorly from a privacy perspective and what works well. You might even add some helpful friction to experience patterns to intentionally slow down your users.

Learn What Not to Do from Anti-Patterns

Anti-patterns aren't the same as deceptive patterns. The AT&T and Bell Labs programmer Andrew Koenig coined the term *anti-pattern* in the mid-1990s. He explains that "An antipattern is just like a pattern, except that instead of a solution, it gives something that looks

---

14 Bösch, "Tales from the Dark Side."

superficially like a solution but isn't one."[15] Think of an anti-pattern as a pattern that was originally devised to solve a problem but was later shown to have some ill side effects. That probable lack of nefarious intent is what distinguishes these from deceptive patterns. Time goes by and when vulnerabilities are discovered within these patterns, better patterns are designed, and the earlier patterns get sent to the dustbin of design history. Regardless, these anti-patterns can still be labeled and presented as a bad example, which means you can learn what *not* to do from them.

That dense thicket of copy you encounter within a privacy policy could be considered an anti-pattern. Privacy policies were instituted, after all, with the intention of ensuring that you're made aware of how companies are handling your data. Ostensibly. Nonetheless, most aren't easy to read. The intention may not always be to deceive, but the language used to convey such important information can make what you're reading seem indecipherable. And if you're looking for very specific information about how your data is used, there's a good chance you'll find it there in a privacy policy. Hopefully. So, by examining a series of privacy patterns, you can certainly pinpoint those characteristics of a policy that aren't working and by inference what might work better.

Some common practices for authentication that were intended to secure your accounts are now considered anti-patterns, too. When experiences asked you some personal questions based on your life and background, for example, that was intended to help secure your information by asking you to present your answers later to confirm your identity. (See Figure 6.17.)

Experts have since concluded, however, that a smart hacker knows where to look to find the answers to many of these questions. Consider, too, that your answers to such questions could change over time. Or you might even forget how you answered them! Consequently, "secret questions" or "identity questions" are now considered security and password anti-patterns.

---

15 Linda Rising, *The Patterns Handbook: Techniques, Strategies, and Applications* (Cambridge University Press, 1998), 387.

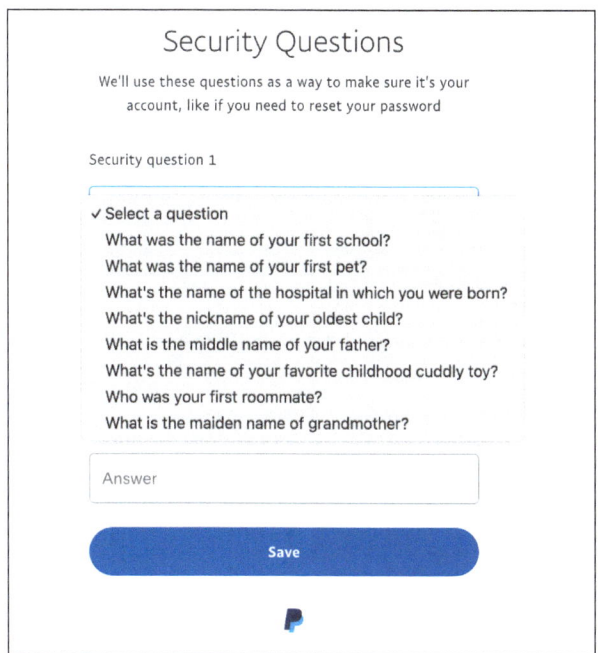

FIGURE 6.17
PayPal still uses security questions to verify your identity in moments such as when you need to reset your password.

Consult Bright Patterns for Better Directions

On the flip side, *bright patterns* (or, sometimes, *fair patterns* or *conscious patterns*) attempt to portray best practices in design patterns that steer clear of deceptive patterns. If you apply bright patterns to privacy, you can imagine cookie banner designs that place an emphasis upon making consent approval or denial equally attractive. They might demonstrate how to publish a palatable, transparent privacy policy. Or they might take a proactive stance by offering to remind you to check your privacy settings in the future as you can in Figure 6.18, an example from Google that could be considered a bright pattern.

In his book *Deceptive Patterns*, Harry Brignull makes a sobering point about so-called "bright patterns," however. "On their own," he says, "bright patterns are really just *yet more educational materials* that appeal to the reader's moral code to do the right thing, which so far hasn't worked." It's a valid point. If you're looking for good examples, you can probably find them readily. The real issue may be convincing people why they shouldn't use deceptive patterns. Still, knowing that this label exists for specific, more ideal design patterns may help you in your research to craft better patterns in the future.

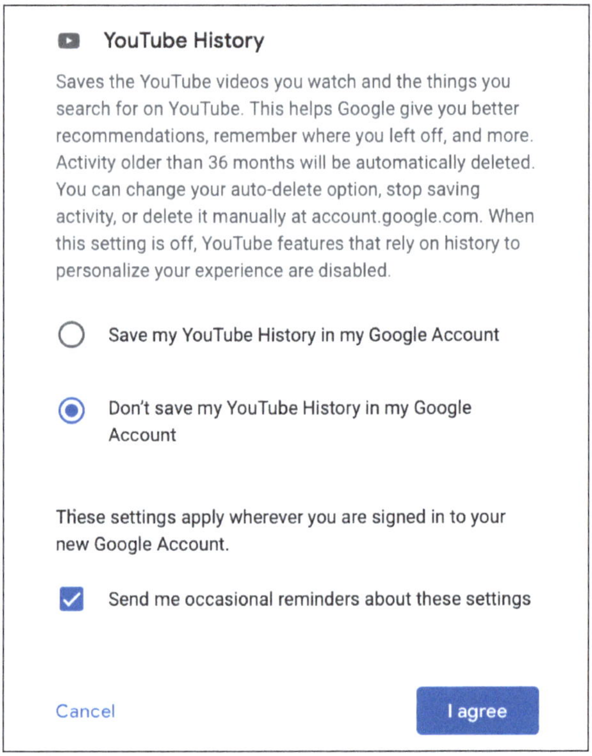

FIGURE 6.18
In this onboarding screen that users see when setting up a new account, Google presents them with the choice to save their YouTube history or not along with a checkbox that, if selected, allows Google to "Send me occasional reminders about these settings."

> **NOTE** **PRIVACY PATTERNS**
>
> Created by UC Berkeley's School of Information, the website **privacypatterns.org** showcases a large set of privacy-specific patterns, explaining the problems with many patterns, as well as their consequences and possible solutions.

Add Helpful Friction

The fact that deceptive patterns often seek to accelerate important decisions that users need to consider carefully indicates that adding friction to user experiences is sometimes desirable or even necessary. For example, if you're about to make a large online payment, you may appreciate an extra step in the process that allows you to confirm that you've selected the right product, entered the right amount, or selected the right payment date. So, too, in those cases where business is tempted to deploy deceptive patterns into an experience, these may prove to be moments where those experiences would benefit from *more* friction, not less.

An unselected checkbox next to a third-party sharing statement means that users have to take a moment to review what they're being asked to do and to consciously decide to select it. (Of course, they may not consider it at all, which is why companies are so tempted to implement these patterns.) However, if a preselected checkbox means that users might accidentally share their personal information, then the solution should be, instead, to keep it unselected and allow them to just skip over it. If the request is important enough and justified, then business should find ways to draw attention to the selection and explain its value. Yes, they should *add* more friction. (See Figure 6.19.)

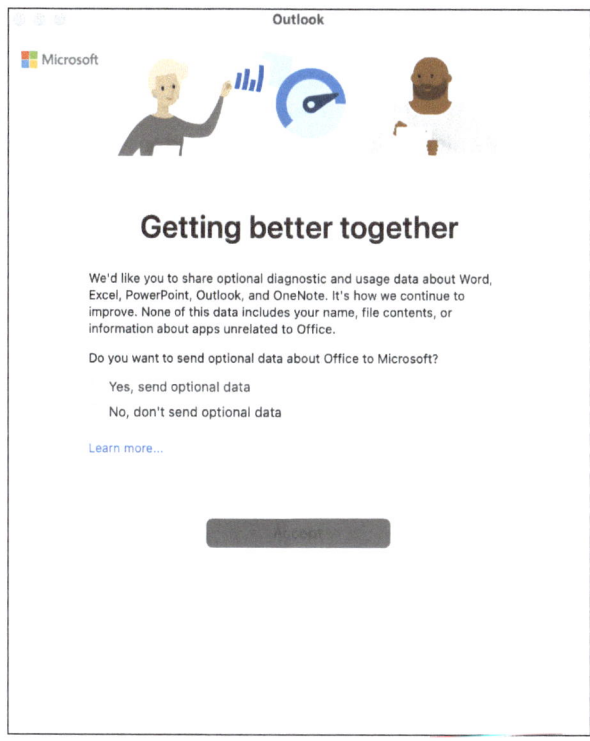

FIGURE 6.19
In this screen that users see when installing Microsoft's Outlook, both radio buttons are left unselected, prompting the user to review the options in more detail and to make a selection before they can continue.

For example, by providing an onboarding screen that explains how data will be used before users continue to fire up an app for the first time, you ensure that they'll be better educated about how this new experience works and more likely to find it trustworthy.

INTERVIEW WITH HARRY BRIGNULL

Harry Brignull is a user experience director who coined the term *dark pattern*, which he later updated to *deceptive pattern*. He provides testimony as an expert witness in lawsuits involving deceptive patterns. He wrote the book *Deceptive Patterns: Exposing the Tricks Tech Companies Use to Control You* and is the founder of deceptive.design.

What if anything specifically first provoked your interest in deceptive patterns?

Back in the early 2000s, I did a Ph.D. in Cognitive Science and when I completed it, I became a user researcher. This was back in the days when user research was a super rare skill that suddenly became highly in demand. The work was repetitive but quite easy—in those days, digital products had pretty poor baseline usability. There were lots of design issues that caused users to become confused and give up, which cost our clients money. It was my job to find those issues by observing users carrying out activities in their products. And when you find and fix usability issues, it's a win for both sides. Users get what they want—to complete their activities without being frustrated or blocked—and clients get to make more money.

But as I did this research, sometimes issues would come up that the client did not want to fix. Initially, I noticed simple things like marketing checkboxes where users had to carefully read the trick language and tick some checkboxes but not others to fully opt out. When I reported this, some clients would say, "Well we're not changing that, it makes us a ton of money, thanks anyway." This would stick in my mind like a painful splinter that didn't fit into the philosophy of design that I'd been taught. If you look back at the history of UX, it was mainly born from the fields of HCI (human-centered interaction) and Human Factors—it was all about understanding user psychology to help people prevent errors. In safety critical systems, it was about preventing catastrophes from happening. But somewhere along the line, this framing of UX transformed into being about influence, persuasion, control, and measurement.

This is what drove me to start campaigning about deceptive patterns. It was born of frustration, and it was an effort to get the industry to face up to what was really happening.

Have you come across a deceptive pattern that undermined people's privacy that stands out for you in particular?

Well, the EU cookie walls are a good example. What people don't realize is that the regulations never asked for the design implementations to be so incredibly frustrating! This was a product of the industry trying to influence consumer behavior by introducing obstacles into the opt-out process.

Are you encouraged by the progress in classifying deceptive patterns and even beginning to regulate them? What more needs to be done?

Around 2015, academic researchers started showing interest in the area, and there was a surge of attention that came from the findings they published. This stimulated a lot more attention from regulators and legislators around the world. It's become quite a big topic, and it is satisfying to see this happening. But the whole point of this work isn't just to study it and talk about it—it's to have an effect on the industry.

What aspect of deceptive patterns do you feel is not getting enough attention currently?

I have to say I'm quite disappointed in the design industry as a whole. If I were to ask someone in design leadership "What are the laws of UX?" they would probably reply with stuff like "Fitts's law?" or "The Law of Proximity?"—design guidelines, principles and that sort of thing. Yet there are lots of real laws—actual legislation that governs our work and that serves to protect consumers. It's mind-boggling that we don't learn it as part of our training. And we spend so much time patting ourselves on the back for caring about users! If we cared about users, we'd learn about the laws that have been written to protect them.

Do you have any concerns around the use of artificial intelligence in creating or enabling deceptive patterns?

Of course. We're currently at the thin end of the wedge, and the first effect we're seeing is the shrinkage of design teams and design budgets. The money that was being spent on design is now being spent on AI initiatives; and also, the new generation of commercial AI tools for designers and researchers are making our activities more efficient, so teams don't need to be as big. Smaller teams have less corporate power, and the candidate-heavy market means that there are 10 people waiting on the sidelines to take your job if you've got any misgivings about your employer's business practices.

continues

INTERVIEW WITH HARRY BRIGNULL (continued)

The other thing that's happened is design commodification. Today, there are so many popular business models that have mature, templated user experiences. These days, it's easy to copy your competitors and with a few tweaks you can end up with a reasonable (if derivative) product. This huge repository of prior art has been swallowed up into the training data of modern multimodal AI tools that make the act of copying so much easier! What's more you don't just get UI mocks from these tools (Bolt, Cursor, Windsurf, etc.)—they give you working front-end code. User research is being chipped away, too. All the administrative work in transcription, summarization, and synthesis has been swallowed up by large language models (LLMs), which means an organization can go through the motions of research—and look like it's doing a good job—without a trained professional really dwelling on the potential negative consequences and taking the time to go out and persuade execs to really care about it.

My point is that we don't have to wait for this stuff to play out—we're already seeing the effect on design teams. For example, there used to be a really healthy industry of professional training—if your team ran into a challenge like accessibility, or design tokens, or some technical aspect of design that your team wasn't strong on, the normal response was to hire an expert to come teach your team how to do it. This has pretty much collapsed over the last 2–3 years because design teams aren't getting the training budgets they used to have.

We can hope that things will rebound somehow, but I think we're in a new era now. I hate to say it, but it feels like the golden age of UX design is over and we're entering a new phase of something new. Let's hope that the next generation of designers and researchers will have enough integrity to see it through.

EXERCISE COLLECTING A CUSTOMER'S CONTACT INFORMATION

Assume that you're working with a company that wants to secure potential customers' email addresses and phone numbers before offering them a discount. How might you design these promotional modules to be transparent about what personal information the company is requesting? And to allow options for what information visitors want to surrender, instead? While still keeping the experience as clear and simple as possible?

Some possible solutions:

- Recommend that the requirements be worded more clearly up front and place both the email and mobile fields on the same screen to create a truly transparent—and more efficient—design.
- Suggest the company offer an *additional* discount on the second screen if the customer adds their phone number, as well. That could prove to be a little annoying since it still places another barrier within a customer's task flow. But, at least, you'll be designing an experience that stays honest and offers something to *both* the company and its customers.

The Takeaway

Fortunately, we human beings are pretty good at pattern recognition. After studying deceptive patterns in some detail, you'll likely experience a pop of recognition when you spot them in the wild now. That's good, since you'll know better how to address them in your own work and how to avoid them in the first place. And you'll become an even better designer in the process.

Here are some resources worth reviewing to learn more about this important aspect of designing for privacy.

Projects

A few different parties have made significant contributions to identifying and codifying deceptive design patterns, including the following:

- **Deceptive Design:** As a team, Harry Brignull, Mark Leiser, Cristiana Santos, and Kosha Doshi work to highlight a growing library of deceptive patterns, as well as a robust "Hall of Shame" that includes many searchable examples—deceptive.design.
- **Dark Patterns Ontology:** Colin M. Gray, Nataliia Bielova, Cristiana Santos, and Thomas Mildner—ontology.darkpatternsresearchandimpact.com
- **"Guidelines 03/2022 on Deceptive Design Patterns in Social Media Platform Interfaces: How to Recognize and Avoid Them":** Adopted by the European Data Protection Board in February 2023, these guidelines focus specifically on avoiding deceptive patterns common to many social media platforms.
- **"Tales from the Dark Side: Privacy Dark Strategies and Privacy Dark Patterns":** A paper presented at the Privacy Enhancing Technologies Symposium in 2016 by Christoph Bösch, Benjamin Erb, Frank Kargl, Henning Kopp, and Stefan Pfattheicher.

Books

- Harry Brignull: *Deceptive Patterns: Exposing the Tricks Tech Companies Use to Control You* (Testimonium Ltd., 2023).
- Raphael Gellert, Hanna Schraffenberger, Cristiana Santos: *Dark Patterns & Manipulative Design: Conceptualising and Systematising a Key Contemporary Phenomenon from a Legal Perspective and Beyond* (Cheltenham, England, Edward Elgar, Spring 2025).

CHAPTER 7

Use Language with Care

Deciphering the Indecipherable	120
Aim for Clarity	123
Keep Content Honest	134
Make Navigating Content Easy	141
The Takeaway	146

Ever curl up by the fire to read a good privacy policy? Me neither. According to a 2019 Pew Study, just 9 percent of Americans say they always read a company's privacy policies before agreeing to them.[1] That's unsurprising considering privacy policies keep ballooning in length. A 2018 study found they average 3,964 words in length, up some 58 percent from the results of a 2008 study.[2] Still, those findings may be better than what Pew found in 2014,[3] when they asked Americans the following true/false question: "When a company posts a privacy policy, it ensures that the company keeps confidential all the information it collects on users?" about 52 percent of respondents said this was true. Of course, it was not.

Although as a designer, you may not always have control over how copy and content is developed within an experience, there are some best practices you can keep in mind to ensure this important privacy content is presented as transparently and effectively as possible.

- Aim for clarity.
- Keep content honest.
- Make navigating content easy.

In some cases, you will be able to design directly to satisfy these best practices. In others, you may at least be able to present these as input or guidance for those working on the content.

Deciphering the Indecipherable

In 2019, after conducting what must have been one of the most grueling research projects you could imagine, the data journalist Kevin Litman-Navarro shared the results of a study in *The New York Times* that examined 150 privacy policies from popular sites and apps.[4]

---

1 Brooke Auxier et al., "Americans' Attitudes and Experiences with Privacy Policies and Laws," Pew Research Center, 15 November 2019, www.pewresearch.org/internet/2019/11/15/americans-attitudes-and-experiences-with-privacy-policies-and-laws

2 Pierre-Nicolas Schwab, "Reading Privacy Policies of the 20 Most-Used Mobile Apps Takes 6h40," *Into the Minds* (blog), 28 May 2018, www.intotheminds.com/blog/en/reading-privacy-policies-of-the-20-most-used-mobile-apps-takes-6h40

3 Aaron Smith, "Half of Online Americans Don't Know What a Privacy Policy Is," Pew Research Center, 4 December 2014, www.pewresearch.org/short-reads/2014/12/04/half-of-americans-dont-know-what-a-privacy-policy-is/

4 Kevin Litman-Navarro, "We Read 150 Privacy Policies. They Were an Incomprehensible Disaster," 12 June 2019, www.nytimes.com/interactive/2019/06/12/opinion/facebook-google-privacy-policies.html

He described what he found as an "incomprehensible disaster." He found issues with most of these policies but described Airbnb's as "particularly inscrutable."

A sample statement from Airbnb's privacy policy at the time read: "This information is necessary for the adequate performance of the contract between you and us and to allow us to comply with our legal obligations."

As anyone who has attempted to read a privacy policy knows, language is often used to intentionally obscure privacy issues and to confuse or even trick users, instead of educating them. As Litman-Navarro found, Airbnb used vague language that "allows for a wide range of interpretation, providing flexibility for Airbnb to defend its data practices in a lawsuit while making it harder for users to understand what is being done with their data." When companies use language like this, they are setting the advantage in favor of their lawyers, not the user.

Litman-Navarro also added an education lens to his findings, finding that most of these policies required a college or professional career level of reading comprehension. Litman-Navarro concluded, "A good privacy policy would help users understand how exposed they are."

In contrast then, Litman-Navarro determined that the BBC had "an unusually readable privacy policy," which he said was written at a middle school level, "in short, declarative sentences, using plain language." (See Figure 7.1.)

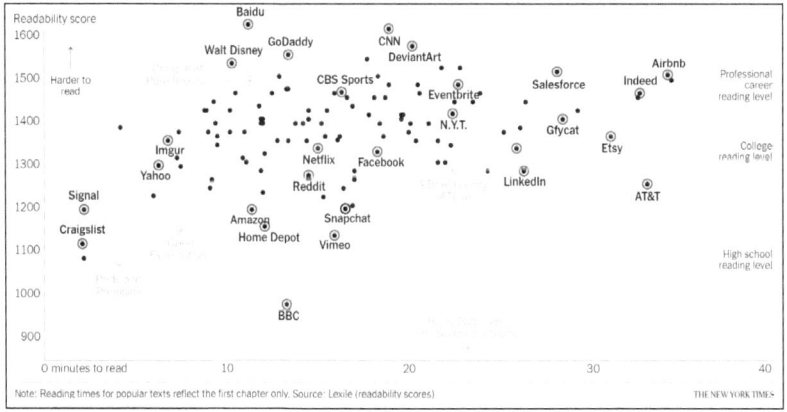

FIGURE 7.1

This scattergram from Kevin Litman-Navarro's study positions platforms' privacy policies according to how difficult they are to read and how long it takes to read them.

Consider Figure 7.2 from the BBC's site to see how it contrasts with so many other site's privacy policies.

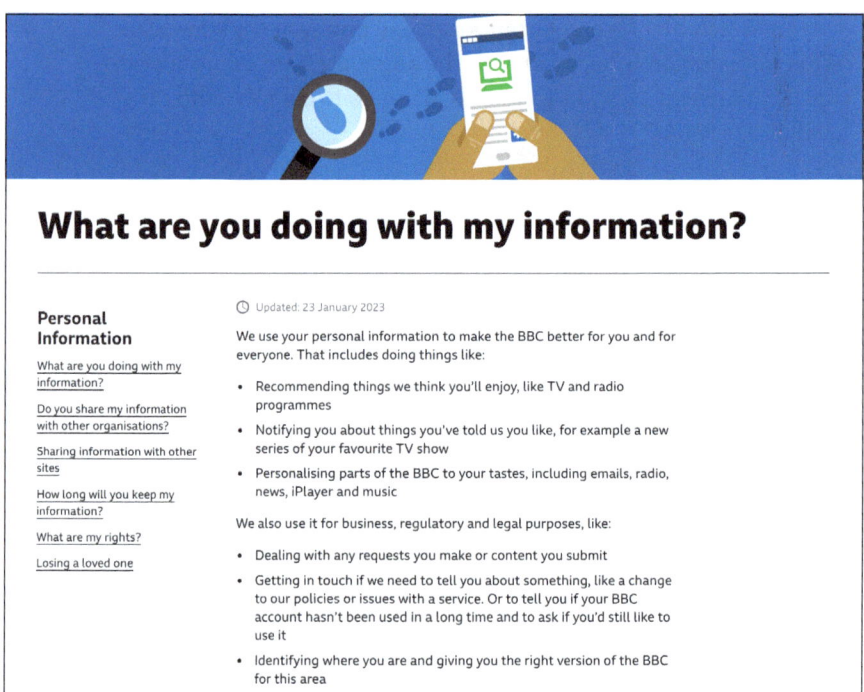

FIGURE 7.2
Note the clear, straightforward language the BBC uses for their privacy policy.

What makes the BBC's policy more digestible?

- **Simpler language:** The copy reads more easily and seems written at a level appropriate for most readers.
- **Information clustering:** The content doesn't look like one endless, scrolling paragraph of legalese. Instead, it's broken down and clustered into meaningful chunks of information.
- **Navigation:** The content has been sorted into categories that are reflected in the navigation to the left of the screen, allowing visitors to skip easily to those sections that concern them most. Since the site is designed responsively, these navigation items appear within a drop-down on the mobile version.
- **Use of bullets:** Listable items are placed in bullets to ensure better scannability.

Whether you're developing a privacy policy or other content relevant to privacy issues, here are some ways you can ensure that privacy-oriented content empowers users.

> **NOTE WINE AND COOKIES**
>
> In an amusing attempt to gauge whether anyone was reading his company's terms and conditions policy, Dan Neidle, the founder of Tax Policy Associates, included the following in a bullet explaining how the company's website used cookies: "We will send a bottle of good wine to the first person to read this."[5] Three months later, someone claimed the gift.

Aim for Clarity

Clarity is a key characteristic for any sort of good writing. If you genuinely hope to communicate ideas effectively and honestly, you'll avoid writing in any way that creates ambiguity or confusion or otherwise creates obstacles to reading comprehension.

Clear, concise writing becomes especially important when writing privacy-related content. Sadly, however, it's this content that's often cast—sometimes quite intentionally—in ways that make it as impenetrable as possible.

"Plain language" may seem like a subjective standard, but increasingly regulations require it. For example, the GDPR not only requires transparency as a broad principle for processing personal data, but it also specifically requires "communication ... relating to processing to the data subject in a concise, transparent, intelligible and easily accessible form, using clear and plain language, in particular for any information addressed specifically to a child."[6]

---

[5] Morning Edition, "Tax Policy Think Tank Founder Was Curious If Disclaimers' Fine Print Was Being Read," NPR, 13 May 2024, www.npr.org/2024/05/13/1250855157/tax-policy-think-tank-founder-was-curious-if-disclaimers-fine-print-was-being-re

[6] "Art. 12 GDPR, Transparent Information, Communication and Modalities for the Exercise of the Rights of the Data Subject," GDPR, https://gdpr-info.eu/art-12-gdpr

Avoid Jargon and Legalese

Good writing should always avoid using jargon, but jargon is often used specifically to obfuscate privacy issues. University of Pennsylvania Professor Joseph Turow surveyed consumers on the topic of digital marketing and privacy. He concluded the following: "[R]esearchers have found that people do not read privacy policies—they're unreadable. They are filled with jargon that is meant to be understandable only to the people writing them, or to people who work in the advertising industry today. Words like 'affiliate': nobody outside of the digital marketing industry knows what that means."[7]

It might be surprising to learn that many lawyers advocate for the better use of language, too. That's right—even lawyers don't like legalese. In a 2023 study, MIT researchers concluded that lawyers write that way, because, well, it's just the way they've always written.[8] In fact, Edward Gibson, an MIT professor of brain and cognitive sciences, concluded, "No matter how we asked the questions, the lawyers overwhelmingly always wanted plain English."[9]

These researchers found that one specific practice contributed the most to making "legalese" so dense and impenetrable, something linguists have termed *center embedding*. This refers to the practice of inserting clauses within clauses. In legal writing, these clauses often turn out to be lengthy legal definitions that are plonked into the middle of already complicated sentences. As Gibson said, "In normal language production, it's not natural to either write like that or to speak like that."[10] The researchers concluded that lawyers may just believe they need to write this way to sound professional, but also that they just get used to reading content formed this way.

This problem of center embedding becomes particularly relevant when you're considering privacy statements or copy in general,

---

7 Aaron Smith, "Half of Online Americans Don't Know What a Privacy Policy Is," Pew Research, 4 December 2014, www.pewresearch.org/short-reads/2014/12/04/half-of-americans-dont-know-what-a-privacy-policy-is/

8 Eric Martínez, Francis Mollica, Edward Gibson, "Even Lawyers Do Not Like Legalese," *Proceedings of the National Academy of Sciences of the United States of America*, 6 June 2023, www.ncbi.nlm.nih.gov/pmc/articles/PMC10266064/

9 Anne Trafton, "Even Lawyers Don't Like Legalese," MIT News, 29 May 2023, https://news.mit.edu/2023/new-study-lawyers-legalese-0529

10 Ibid.

which affects people's privacy. Readers really need to grasp the meaning of terms to properly process this information (see Figures 7.3 and 7.4). One solution is to define terms separately and, preferably, up-front before users launch into the full text. Avoiding center embedding also reduces the length of sentences, an important way of making legalese more digestible.

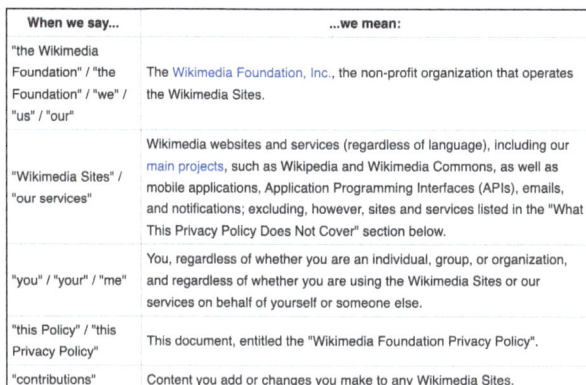

FIGURE 7.3
Wikimedia provides an explanation of terms before reading their full privacy policy and also provides a link to a more detailed "Glossary of Key Terms."

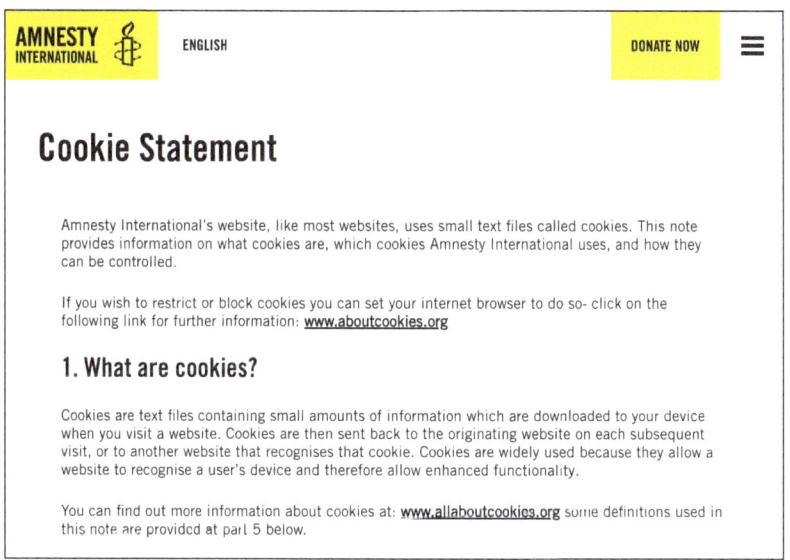

FIGURE 7.4
Amnesty International dedicates a page to defining cookies and explaining their function in plain language.

Legalese is most often found in privacy policies and terms-of-use sections of websites. You might also find them labeled *Legal* or *Terms and Conditions* or something else. Sometimes, these areas may be combined. Sometimes, they may consist of several pages of privacy content. Some sites—though very few—even include a privacy portal, a dedicated space with its own navigation, multiple pages of relevant content, privacy-oriented tools and settings, maybe even with its own URL or subdomain. Often, links to this important information can be found only in the website's footer. Additionally, legalese may appear in disclaimers or in what we often call the *small print*.

Of course, as a designer—or a copywriter—you likely won't be writing a privacy policy. The company's legal department will be. And if you're working with a larger company, it may be especially difficult for you to influence how policy content is handled. But that doesn't mean you can't resist the use of jargon and opaque language elsewhere in the experience, especially when it pertains to privacy, and it doesn't mean you can't position yourself to give some feedback on how legal content is *presented* from an experience perspective. For example, you may have more luck helping in the following ways:

- Offering advice on the formatting, scannability, and navigation of the content, which can help users better consume it.
- Including a summary of key takeaways up front before the legalese takes over.
- Weighing in on how disclaimers and other small-print items are handled elsewhere, which often fall prey to legalese, too.

A WORD ON FONT SIZE

Accessibility guidelines typically recommend at least 12-point (16-pixel) fonts for website copy, but it's not unusual to find sites using much smaller fonts for small print or disclaimers, while using more accessible copy everywhere else (Figure 7.5). Accessibility experts recommend never using a font size of 9 points (12 pixels) or less, as these sizes prove very difficult for most people to read—and particularly difficult for anyone with vision problems.

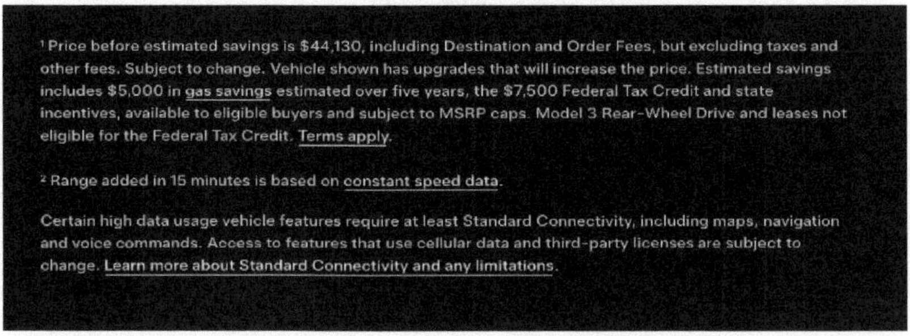

> [1] Price before estimated savings is $44,130, including Destination and Order Fees, but excluding taxes and other fees. Subject to change. Vehicle shown has upgrades that will increase the price. Estimated savings includes $5,000 in gas savings estimated over five years, the $7,500 Federal Tax Credit and state incentives, available to eligible buyers and subject to MSRP caps. Model 3 Rear-Wheel Drive and leases not eligible for the Federal Tax Credit. Terms apply.
>
> [2] Range added in 15 minutes is based on constant speed data.
>
> Certain high data usage vehicle features require at least Standard Connectivity, including maps, navigation and voice commands. Access to features that use cellular data and third-party licenses are subject to change. Learn more about Standard Connectivity and any limitations.

FIGURE 7.5
Like many automotive manufacturers, Tesla uses a much smaller font size to display disclaimers, which may prove difficult for many visitors to read.

Fonts should also be used with relative size values, rather than fixed, so users can resize them, if they need to. Specifically, the Web Content Accessibility Guidelines (WCAG) recommend that all text be scalable up to 200 percent.[11] Although neither the WCAG nor the Americans with Disabilities Act (ADA) recommend an "official" minimum font size, making text size inaccessible could still land a company in legal trouble.

---

11 "Resize Text," World Wide Web Consortium (W3C), www.w3.org/TR/UNDERSTANDING-WCAG20/visual-audio-contrast-scale.html

Some Good Examples

Let's look at a couple of more helpful examples for how to handle legal content. First, consider how Bandcamp presents their Terms of Use. First, note how they include a summary of "the top three things a visitor to this page might be wondering." (See Figure 7.6.)

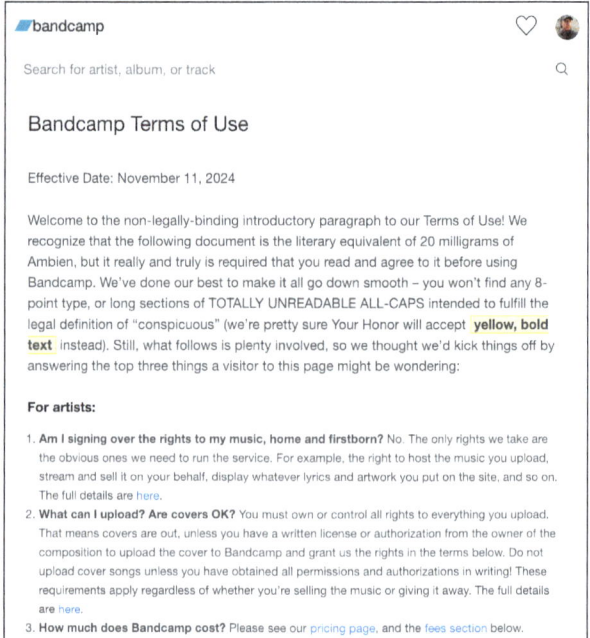

FIGURE 7.6
Bandcamp uses straighforward even humorous language within their Terms of Use.

Not only does Bandcamp highlight specific information important for both artists and users, drawing attention to it with numbering and bold text, but their copy is written with a sense of humor. Look how they highlight the importance of reviewing this content:

> We recognize that the following document is the literary equivalent of 20 milligrams of Ambien, but it really and truly is required that you read and agree to it before using Bandcamp.

See? Legal language needn't always be so boring.

Not every company can write their copy with the same voice as Bandcamp's, but every company can put careful thought into how they present and summarize this content—even if it does mean negotiating with the legal department. Understandably, some designers throw up their hands when faced with restrictions from legal. Instead of giving up, however, see if you can think outside of the box: Find ways to present that content you have little control over in ways which are, at least, more digestible. (And, hopefully, over time, legal departments will feel pressured to use language that's written to an appropriate reading level and less weighted toward intentional obscurantism. Or in plain English: Written so it's easier to read and less sneaky.)

The insurance company Lemonade made a name for itself by making it easy to sign up for more affordable home and contents insurance, but they also built a reputation for explaining their privacy terms clearly and transparently. Lemonade frames their privacy policy as a "Privacy Pledge" that explains at length what the company does with your data (see Figure 7.7). It reads more like an article than a policy and at a pretty approachable level. The opening sentence begins with a plain declaration that any company serious about data privacy should have no problem embracing:

> In light of growing concerns about companies selling customers' private data, we thought this would be a good opportunity to tell you about what types of data we collect, why we collect it, and what we do with it.

FIGURE 7.7
The opening paragraphs of Lemonade's Data Privacy Pledge directly address their customers' potential concerns around the use and sale of their data.

REALITY CHECK

Somewhat ironically, an ostensibly privacy-forward company like Lemonade can still fall prey to privacy lapses: In 2022, Lemonade settled a lawsuit that alleged that the company had collected biometric data from video claims in a way that ran afoul of the Illinois' Biometric Information Privacy Act. Lemonade says they no longer collect biometric information from video submissions (see Figure 7.8).

FIGURE 7.8
Lemonade retracted a thread from Twitter after backlash to the news that they collected biometric data from submitted videos.

Earlier, in 2021, Lemonade had taken to Twitter to boast that they had launched cost-saving AI technology to review video "for signs of fraud" potentially via "non-verbal cues." Their thread left a bad taste in people's mouths with readers criticizing the company for relegating this task to potentially biased algorithms. People worried that decisions on their claims might be based on their skin color. Eventually, Lemonade deleted what they termed "the awful thread," adding "Our systems don't evaluate claims based on background, gender, appearance, skin tone, disability, or any physical characteristic." And they declared, "We never let AI auto-decline claims." Still, some damage was done. Full disclosure? I'm a Lemonade customer, and their service has worked well for me when I needed it.

That will be a common theme of this book: You'll see the same companies provide good examples for how to handle privacy matters, only to find illustrations of the same companies being guilty of bad privacy practices elsewhere. Some of them will be companies you interact with as a customer and maybe even like. It's complicated.

Write for the Right Reading Level

Avoiding jargon and legalese is good, but it's not enough if you hope to communicate important privacy information effectively. Whether addressing the usual suspects like privacy policies or terms and conditions, or more visible content like marketing copy, it's important to write to the appropriate audience level (see Figure 7.9).

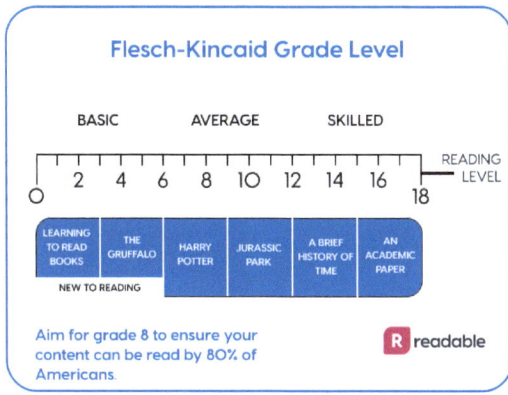

FIGURE 7.9
The Flesch-Kincaid Grade Level test is the best-known tool for measuring how difficult text is to read and understand.

According to the National Literacy Institute, 54 percent of adult Americans read below a 6th-grade level.[12] And 21 percent of U.S. adults are illiterate. The Content Marketing Institute recommends writing at the 8th-grade level to ensure a wider audience and starting at the 10th grade level for books and white papers.[13] So, if you're not writing to the level of approximately a 14-year-old, you're likely making it difficult for readers to digest important information. If you're not sure what level you're writing at, you can try dropping some of your copy into a tool like the Hemingway Editor, which helps determine what your is (see Figure 7. 10). (Be careful, of course, not to paste content into online tools, which may include personal data or proprietary information.)

---

[12] "Literacy Statistics 2024–2025 (Where We Are Now)," National Literacy Institute, www.thenationalliteracyinstitute.com/post/literacy-statistics-2024-2025-where-we-are-now

[13] Steve Linney, "Content Readability: A Primer," 15 October 2018, Content Marketing Institute, https://contentmarketinginstitute.com/articles/boost-content-readability

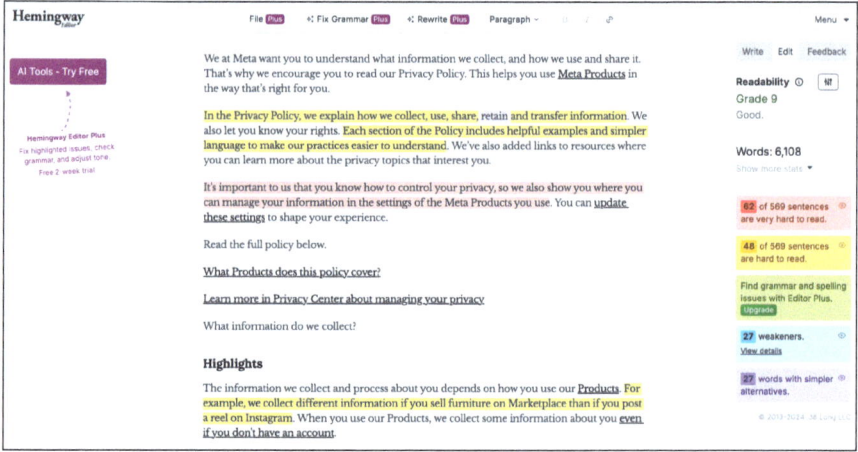

FIGURE 7.10
Hemingway Editor results for Facebook's privacy policy as accessed October 1, 2024.

In early 2022, Facebook launched a rewritten privacy policy that aimed for a more accessible reading level. Unfortunately, the new policy also grew to three times its original length, clocking in at some 12,000 words.

Curious how Apple might perform? At the time of this writing, Hemingway Editor gave Apple's iCloud Terms & Conditions a "Poor" rating, placing that policy's copy at the "Post-graduate" reading level. As of 2018, Barclays analysts determined iCloud had 850 million users globally—a number that must have grown significantly since then.[14] Interestingly enough, if you drop detailed copy explaining updates to iOS 18 into the Hemingway Editor, you'll find the score for Apple's copy moves to "OK" and "Grade 11." That copy was probably written by a marketing team, not the legal department, but the result is that Apple wrote this 1800+-word explanation of their updated iOS at a level that proves much more approachable than their privacy policy. Unsurprising, perhaps, but an excellent example of how companies treat privacy content.

---

14 Jordan Novet, "The Case for Apple to Sell a Version of iCloud for Work," CNBC, 11 February 2018, www.cnbc.com/2018/02/11/apple-could-sell-icloud-for-the-enterprise-barclays-says.html

Of course, there will be exceptions to writing at an 8th grade level. If you have carefully-crafted personas reflecting a particular audience, reference them to determine if an experience's reading level should vary from that recommended range.

You've likely reached this conclusion, too. If you write to the appropriate comprehension level, it's not only going to help to communicate privacy matters clearly, but it will also communicate a company's entire mission and their products or services more effectively.

A Good Example: #SelfCare

The use of clear language to explain privacy information matters everywhere that touches on privacy at all. That might mean, for example, highlighting features affecting user privacy within the first moments of a user's initial engagement with an experience. In fact, regulations may require users' consent before companies can collect such data.

#SelfCare is an app, shown in Figure 7.11, which the creators say focuses on "helping develop a new model for technology, our relationships with our phones, ourselves, and each other." The app uses approachable language during onboarding to explain how your personal data is used when you're setting up the app.

FIGURE 7.11
#SelfCare uses clear, informal language to explain and gain consent for data collection during the app's onboarding process.

Note the copy there: "We currently use Unity for crash reports and in-app purchases. This data collection is totally optional. It helps us catch bugs and blips."

Then #SelfCare provides two clear calls to action (or CTAs): You can select "I'd like to opt in" or "I'd like to opt out" before proceeding to use the app. If you opt out, they explain how to proceed to opt out and provide an additional message confirming that you have.

This example also illustrates the value of providing contextual explanations to highlight privacy matters within just the right moment of a user experience. More on that in Chapter 8, "Provide Tools for Enabling Privacy," when we discuss onboarding and just-in-time alerts.

Keep Content Honest

Honesty is the best policy—particularly when developing privacy content. Honesty enables trust, and trust is paramount if you expect people to share their personal information. Smart companies spend a lot of time developing trust with people, knowing that the loss of that trust can be catastrophic to their business. If a company questions the return on their investment for, say, developing easily read, transparent privacy content, you can remind them of the tremendous value of trust.

When you make the effort to ensure that the way you write privacy-oriented content is straightforward rather than slippery, and it's transparent rather than opaque, you gain that trust. When users feel they don't grasp the true meaning of something despite reading it through multiple times or when they conclude something they find is simply untrue, you lose their trust.

Avoid Deceptive Copywriting

I arrived at the term *desperation notifications* to describe alerts proliferating on social media platforms that use deceptive language to trick users into engagement. Platforms like Facebook, Instagram, LinkedIn, and Twitter (now X) have taken to using this bait-and-switch deceptive pattern that uses copy to seize your attention (see Figure 7.12).

FIGURE 7.12

Much as I enjoy hearing from my LinkedIn friend, he wasn't really addressing me directly.

Remember the LinkedIn deceptive pattern example from Chapter 6, "Avoid Deceptive Patterns"? LinkedIn suggested you import all your contacts to "Stay in Touch," while further reading revealed that LinkedIn would spam your entire contact list at least three times per person if you selected "Add to Network."

Companies use convoluted language like this to trick users into surrendering their private information, too. Prototypr founder Graeme Fulton has highlighted how companies use euphemisms to obscure privacy issues so that "tracking" becomes "personalized ads," for example.[15] Apple highlighted this fact when they debuted their "App Tracking Transparency." Now, when an app suggests that you allow personalized ads, Apple bluntly asks, "Allow Tracking" or "Ask Not Track," highlighting that the company wants to "track you across apps and websites owned by other companies." (See Figure 7.13.)

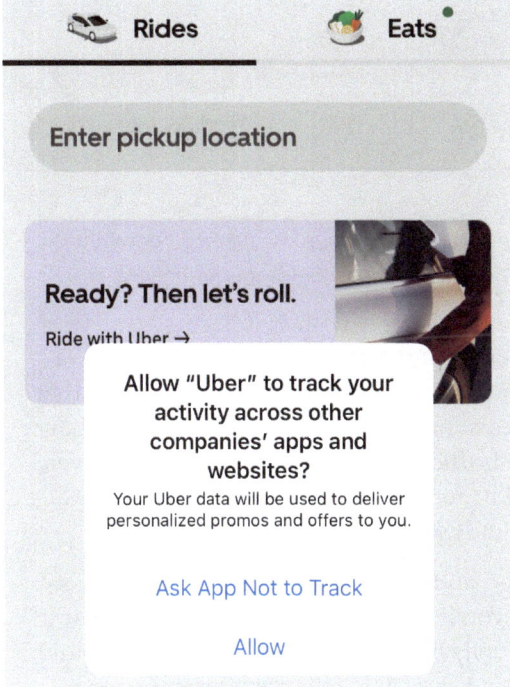

FIGURE 7.13 Apple's feature prompts you to respond whether you'll let Uber track you across the internet or not. This wording may prompt you to wonder, "If I'm only *asking*, does that mean the app may track me anyway?" In fact, this selection only stops the app from accessing your phone or tablet's advertising identifier. The app may still have other ways of tracking you.

---

15 Graeme Fulton, "We Value Your Privacy (at About $0.50): Dark Patterns in UI Copy," Prototypr, 2 February 2021, https://prototypr.io/post/we-value-your-privacy-at-about-0-50-dark-patterns-in-ui-copy

You'll see cookie banners as examples quite a few times in this book since they're a common, prominent feature that companies have implemented to comply with the GDPR. Cookie banners often use deceptive language to entice people into sharing the maximum amount of their data.

Take, for example, common copy like the following in Figure 7.14 that says, "We use cookies for the best experience on our website."

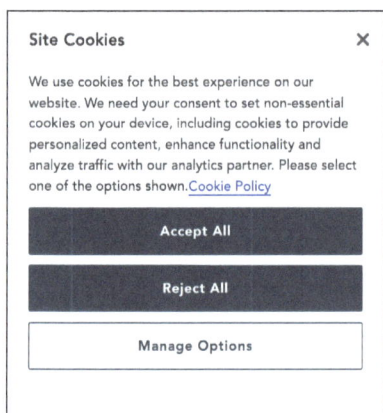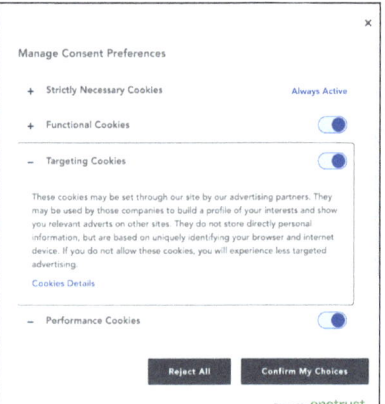

FIGURE 7.14
A cookie banner from the site for a law firm, plus the more detailed Targeting Cookies information you can find by digging further.

Here it's used within a cookie banner on a law firm's website. "We need your consent to set non-essential cookies on your device," they explain, "including cookies to provide personalized content, enhance functionality and analyze traffic with our analytics partner." Given the little copy there, you'd be forgiven for thinking this means that your cookies will be used only for this website's experience.

Click the "Manage Options" button, however, and open the "Targeting Cookies" drawer, and you'll find those cookies belong to their advertising partners and "may be used by those companies to build a profile of your interests and show you relevant adverts on other sites." No doubt, this practice is well-known among those of you working within the web industry. Still, it's not reliably understood by consumers that, when they read that a company uses "cookies for the best experience" on their site, the company also means "we sell your data to enable personalized advertising across the internet."

When you see an expression like "best experience," ask yourself, "Best experience for whom?"

Resist using what Graeme Fulton calls *humbug headers*, too.[16] He describes these as the "use of friendly headings to deflect negative things." He refers to an example from Twitter where the heading for a module says, "You're in Control," but the copy and CTAs are calculated to convince you to "Turn on Personalized Ads."

Keep CTAs and Other Labels Clear and Honest, Too

Even something as simple as a label can be deployed to obscure privacy issues. Labels should always be short and intuitive, of course, because they often serve as navigation or signposts for experiences, helping users to know whether or not they can proceed with confidence. But it's also important to keep labels and calls to action honest. They shouldn't use deceptive language, as shown in Figure 7.15, to trick users into an action.

In this example, a cookie banner from the Safety Services Company asks, "Who doesn't love cookies?" (Another humbug header!) After some brief explanation saying that allowing cookies helps "enhance your browsing experience, serve personalized content, and analyze our traffic," the module offers CTAs to "Learn more" or "I want the BEST experience!" This wording skews heavily toward users sharing the maximum amount of their personal information. Clicking through from the "Learn more" CTA allows users to deny functional cookies, as well as cookies for analytics, performance, advertising, and "uncategorized" cookies. Few visitors will likely click through to discover that, however.

Patterns like these are devised to follow the letter of privacy regulations but not the spirit. We can likely assume they will eventually be banned, and that stricter, more explicit requirements will be added to provide an immediate "reject cookies" option. (In fact, such patterns are already banned under the GDPR and California's privacy law, as well as those of other states.)

---

16 Fulton, "We Value Your Privacy (At About $0.50)."

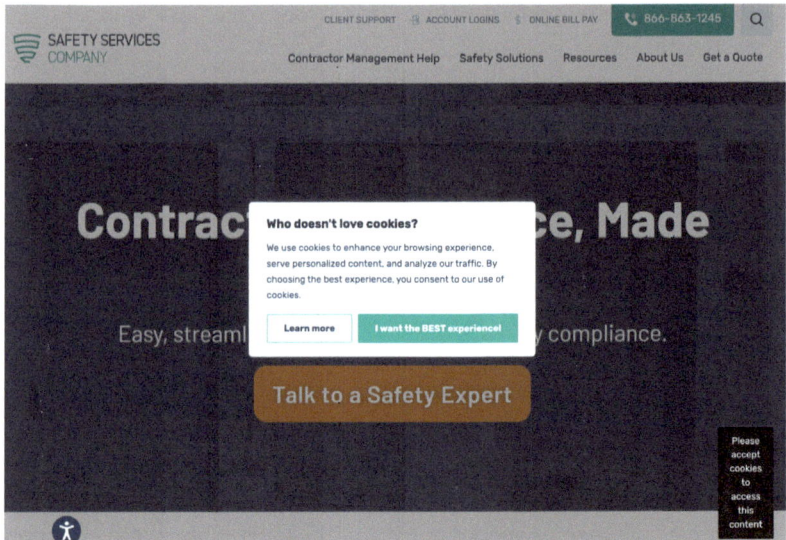

FIGURE 7.15
This banner on the Safety Services Company site uses manipulative language in an effort to convince users to accept cookies. Another message on the screen suggests that visitors can't see the content without accepting cookies, although it's possible to click through via the "Learn more" button to reject five different types of cookies.

Currently, many cookie banners utilize the wording "Allow necessary cookies" beside a button that closes the banner. But what are "necessary cookies"? The GDPR defines "strictly necessary cookies" as those cookies that "are essential for you to browse the website and use its features, such as accessing secure areas of the site."[17] They're generally first-party, session-based cookies, such as the ones that keep your items in an online shopping cart. The GDPR doesn't require consent for these cookies but says companies should explain what they do and why they're needed. In short, any cookies dubbed "necessary" should be required by an experience for it to function properly.

Still, not all companies adhere to the GDPR's definition—nor do they need to if they operate outside of the EU. However, if they

---

17 "Cookies, the GDPR, and the ePrivacy Directive," GDPR.EU, https://gdpr.eu/cookies

deceptively classify advertising or analytics cookies as "necessary," those companies are noncompliant with GDPR.

Several of these examples feature cookie banners, but keep in mind the same issues with deceptive or opaque copy, labels, and CTAs apply elsewhere. They are among the building blocks for deceptive patterns, too.

A good challenge would be to take this practice of honest copy a step further by using language to educate instead of obscure. Instead, you could create content that educates people as to why they should pay attention to how their personal information is being used.

EXAMPLE: *THE SUN*

Let's look at how the British tabloid paper *The Sun* has used leading language, humbug headers, and sneaky CTAs to push users toward accepting cookies (see Figure 7.16).

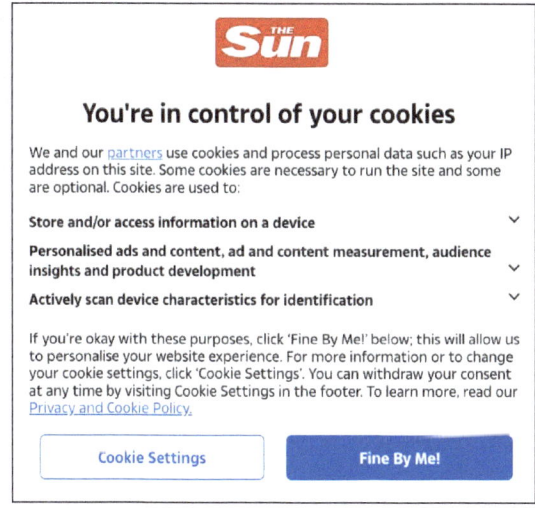

FIGURE 7.16
The Sun's cookie banner leans heavily toward getting users to share the maximum amount of their personal data.

In this 2023 example, *The Sun* used the "You're in control" language, while tilting the copy and CTAs heavily toward accepting all cookies. As is still quite common, they offered no equally placed "Reject Cookies" button.

continues

EXAMPLE: *THE SUN* (continued)

In an updated example from *The Sun*, the paper dropped the "Fine by Me!" language, but now suggests that users can "Pay to Reject" cookies (see Figure 7.17). They also updated the button colors to give them equal weight, and you can now find in the fine print that they share your information with 371 partners, although the copy is not worded to make that clear.

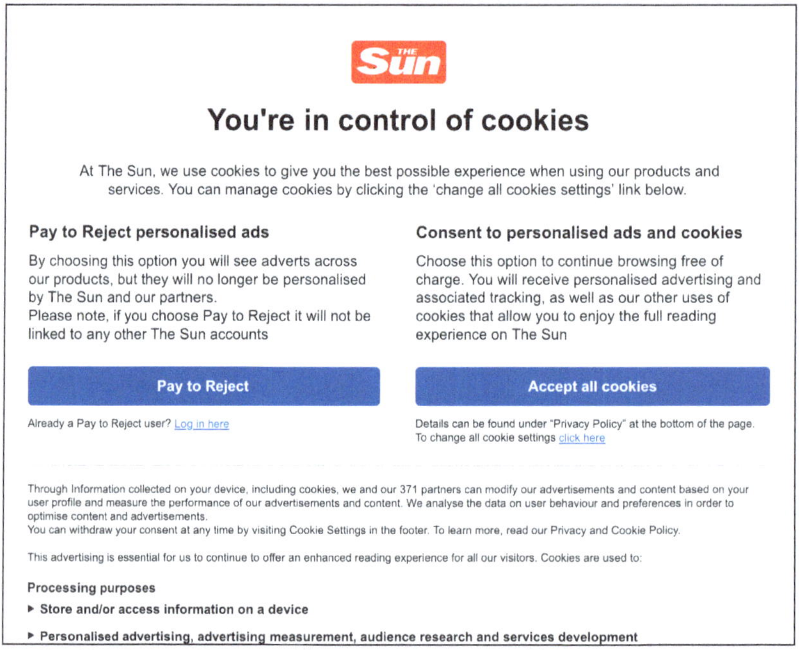

FIGURE 7.17
The updated cookie banner for *The Sun* still downplays privacy options but also suggests users pay to avoid personalized ads.

Still, visitors might not notice that you can still reject cookies, although you would have to click a link embedded in copy appearing in smaller font that says "To change all cookie settings, click here." Rather than improve the privacy experience for their users, *The Sun* found a new way to monetize it via a deceptive pattern.

Make Navigating Content Easy

You can also ensure that privacy-oriented content is more easily understood by using established UX practices that ensure that it has a clear, navigable structure.

We humans don't do well processing and differentiating big blocks of solid content. It's like trying to swallow a handful of huge horse pills without any water. Fortunately, when you apply the following best practices to privacy-oriented content, it not only helps ensure that people can digest this content more easily, but it also shows that you care about being clear and transparent when explaining these important matters.

These best practices focus on the following topics:

- Information clustering and hierarchy
- Navigation
- Scannability

Information Clustering and Hierarchy

The terms *information clustering* and *information hierarchy* serve to describe how you can analyze content to group related information into buckets or categories (information clustering) and then establish a hierarchy for presenting that information (information hierarchy). These practices are imperative for organizing privacy information and then determining what may need to be highlighted or given special attention.

When you group related information together on a screen, you're engaging in information clustering. This exercise is valuable for ensuring that users can find everything about a particular topic in one, clearly labeled place. When you complete a card sort, sticky note, or affinity mapping exercise, you're working to cluster information in meaningful ways, too.

Once you have these buckets of information, you might place them in a high-level sketch or blocking diagram to depict a proposed hierarchy you could review with stakeholders to confirm the priority of how information should be displayed. These two exercises should be standard for any designer placing content on a screen, but you'll often encounter experiences where not enough thought has gone into how the content is organized and prioritized.

Remember to summarize key points. Assuming that the content has been thoughtfully organized and prioritized, you could create a more helpful privacy policy by pulling out some of that content into an overview to highlight, indicating what's most important for users to review if they're not going to read the whole policy. In Figure 7.18, the social media app Bluesky provides an overview of specific elements of their privacy policy they think users may be looking for. This overview precedes the full details of their policy that continues beneath that.

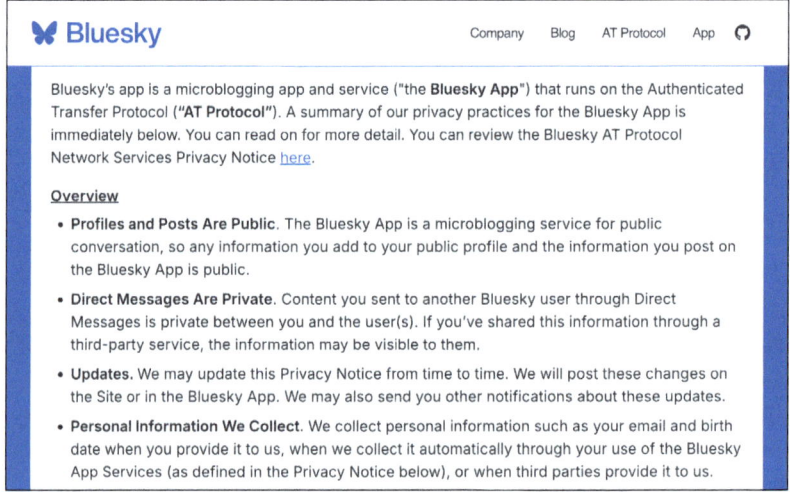

FIGURE 7.18
The Overview section of Bluesky's privacy policy helpfully bubbles up key points for users to pay attention to.

Navigation

Once you've organized your content, consider whether, due to the amount of information, you need to add some form of navigation—for two reasons:

- So users can see a high-level outline of what this privacy content includes.
- So they can skip to topics that prove important to them.

You can decide what sort of navigation to include, based upon the amount of content (see Figure 7.19). Your options might include a left nav, a sticky nav, a drop-down, anchor links, or even a landing page with modules pointing to separate pages for different topics. You'll want to consider how any navigation patterns will work most effectively within a mobile experience, too.

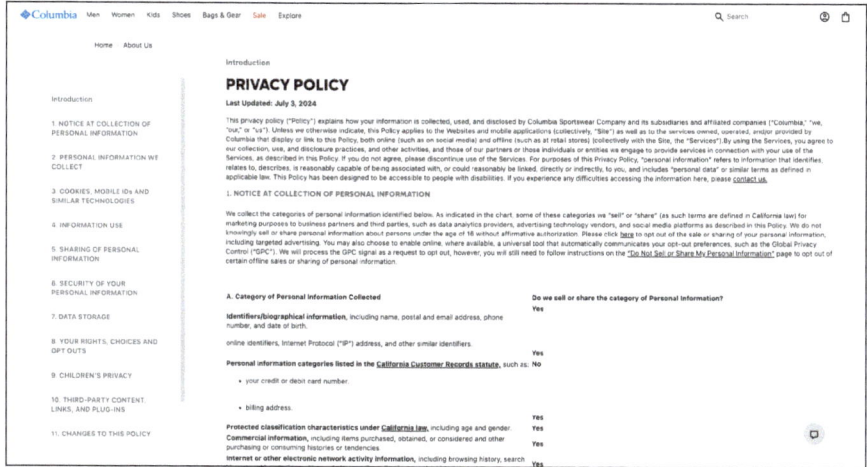

FIGURE 7.19
Note how Columbia has incorporated navigation into their privacy policy. Also note that the portion of the table included indicates that Columbia sells or shares a few different types of personal information.

Scannability

You can still organize your content further, however. Rather than just dumping a huge block of copy onto a page, you can improve its scannability, so that readers can find what they're looking for more easily.

- Headings are an obvious, if often overlooked, way to improve scannability, especially in privacy policies. Pay careful attention to the wording of headings, so they can be understood easily. Keep them short and intuitive to ensure that readers can quickly decide whether a block of content is something they can safely skip—or something they need to pay particular attention to. Placing them in bold and in a larger font ensures that they catch the eye, too, and changing the size of the font for different headings reinforces the content hierarchy.
- Bullets act to improve scannability, too. Don't drop a list of important items, such as technical definitions or parties someone's data is being shared with, into a fat paragraph of

undifferentiated text. Instead, bullet the items out, so they can be reviewed quickly. Of course, you could number the items, too.

Here the Wikimedia Foundation uses both bold sentences and bullets to ensure that the content within the summary of their privacy policy is scannable (see Figure 7.20).

FIGURE 7.20
Not only does the Wikimedia Foundation use bullets and bold text in this summary of their separate privacy policy, but it's written at a very approachable reading level.

- Tables can highlight and explain privacy material effectively. Here, for example in Figure 7.21, Lemonade lists what types of information the company collects, why they collect it, and who they share it with.

And in Figure 7.22, Amnesty International utilizes a table to show why cookies are used and how long they take to expire.

These are common formatting and organizational elements for ensuring that content is easily scanned and consumed, yet they're often absent within impenetrable blocks of privacy content—and even within marketing copy that may touch on privacy issues. Taking care to apply structure to content not only ensures that users can navigate quickly to the information they need most, but it also increases their confidence that your experience is trustworthy and transparent.

FIGURE 7.21

This table from Lemonade's site neatly explains how different types of data are used by the company.

FIGURE 7.22

This table on Amnesty International's site ensures scannability and comparison, too.

USE LANGUAGE WITH CARE   145

In some cases, you may need to employ a little progressive disclosure to display content to avoid overwhelming readers with content but still allow them to dig deeper on a topic important to them. You can allow for presenting more content within accordions or behind calls to action to reveal more content. Just be sure that the labels or other signposts for this content are intuitive and that users don't find important settings hidden behind these disclosure mechanisms.

> **EXERCISE** RETURN TO AIRBNB
>
> Let's return to Airbnb. Have they improved their privacy policy since that *New York Times* study was conducted? Take a look at their site now and see what you find.
>
> Some questions you might ask:
>
> - Is the copy more easily understood?
> - Do they avoid using jargon and legalese?
> - Does the content need and include any form of navigation?
> - Could the content formatting be improved to make the content more digestible?
> - Is it scannable? Does it use headings and bullets effectively?
> - Are headings and labels intuitive?

The Takeaway

I've used privacy policies as an example throughout this chapter, presuming they're a necessary evil, but some experts have reached a different conclusion. Tech journalist Geoffrey A. Fowler suggests that we get rid of them altogether: "We the users shouldn't be expected to read and consent to privacy policies. Instead, let's use the law and technology to give us real privacy choices."[18]

There's little chance, however, that you'll have the freedom to tell your company or client that they should just ditch their privacy policy. Keep these guidelines in mind then, not only to improve the language and presentation of privacy policies, but also any other content that touches on privacy to ensure that consumers can be confident they understand how their personal information is being used.

---

18 Geoffrey A. Fowler, "I Tried to Read All My App Privacy Policies. It Was 1 Million Words," *The Washington Post*, 31 May 2022, www.washingtonpost.com/technology/2022/05/31/abolish-privacy-policies.

CHAPTER 8

Provide Tools for Enabling Privacy

Make Privacy Tools a Priority	148
Ensure That Privacy Features Are Easily Discoverable	154
Follow Best Practices for Privacy Features	161
Remind Users of Privacy Features	171
Never Change Privacy Settings Without Letting Users Know	174
The Takeaway	178

The more personal the data that a business collects from individuals, the greater the responsibility it has to provide tools and settings for those individuals to update, throttle, deny, or otherwise control the flow of their personal information.

If you're working with a large organization that handles a lot of personal information, you might even advocate for a privacy portal, microsite, or toolbox to highlight these features. To ensure that you're providing users with the right tools for maintaining their privacy, consider the following:

- Make privacy tools a priority—on screens *and* in your backlog.
- Ensure that privacy features are easily discoverable.
- Follow best practices for privacy settings.
- Remind users about these privacy features.
- Never change privacy settings without letting users know.

> **NOTE A CAVEAT: IT'S COMPLICATED**
>
> Among the examples that follow, you'll find tools created by some of the same companies you learned earlier were guilty of some pretty grave privacy violations. As we concluded with the renter's insurance company Lemonade in the previous chapter, such is the complicated state of business. Many of these companies are so large that it's possible you'll find people making a good faith effort to create privacy tools in one place, while others devise the means to undermine your privacy in another. As you'll see, sometimes these dynamics even compete within the same screen. Consider this caveat then: The examples shown here—even the positive ones—should not be considered endorsements (or, necessarily, condemnations) of the companies that created them.

Make Privacy Tools a Priority

You've noted by now that many website cookie patterns are presented as if they're assisting users with their privacy but often trick them into accepting cookies instead, burying the important details that explain what data is being shared and with whom. That's emblematic of how many companies, more broadly, make a show of providing a suite of important privacy-related tools or settings but still default to making people's data public and squirrel those helpful tools away deep within their experiences where users are unlikely to dig for them.

It's a joy then to work with companies that do value privacy to the extent that they're driven to create privacy tools and to promote them. You can play an instrumental role, too, in ensuring these tools become more than just suggested features lingering in a backlog, which never seem to struggle their way into a design sprint. That may mean instigating a cultural shift in priorities toward offering privacy features in the first place, as well as giving them visible priority within the experience.

Many larger platforms, especially social media apps, offer a dedicated group of privacy settings with their broader account settings. Some even offer a stand-alone "privacy portal" to gather privacy content and features in one place. Some companies also offer privacy checkups and periodic reminders to check your settings. The more features you offer your users to give them control over their personal information, the more trust and confidence you will build with them.

Privacy Settings

The most common way you'll find privacy-specific features is within a website or app's settings. You may find these presented very simply as a single setting or two that you can adjust to restrict, for example, who sees your profile or can find you. Or you may find a whole section or screen filled with a daunting array of different settings. The latter tends to apply more when users are sharing a lot of personal information, especially on a social media platform, where they need fine-grained control. On the other hand, you might conclude, in some cases, such a bewildering array of individual settings may appear with the intention of making it *more* onerous and time-consuming to ensure that your account remains truly private.

Unsurprisingly, Facebook and Google have dedicated areas for these privacy settings, not just a single screen. If you visit your Google account, you'll find the privacy settings are prominently promoted there, in addition to appearing in the navigation (see Figure 8.1). Google also promotes privacy and security recommendations there.

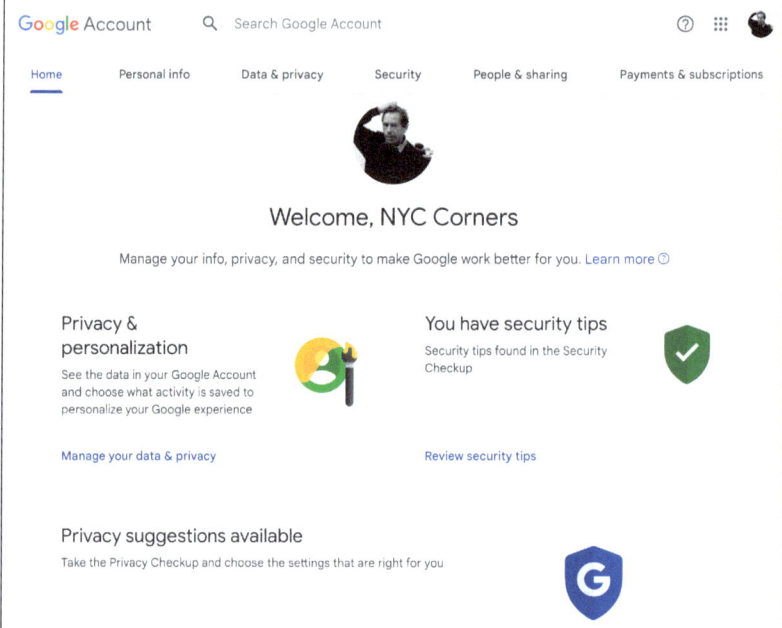

FIGURE 8.1

The home screen for Google's account settings promotes reviewing your privacy and security settings and recommendations. Users also find a section devoted to "data and privacy."

If you click into these screens, you'll find a long list of different ways to modulate your privacy settings. You'd likely expect to find settings for your search history and web browsing there, and you will, presuming you're using Google's search and their Chrome browser. However, you're likely also using Google Maps and YouTube. You might be using Google Drive or Google Fit or any number of other Google apps or third-party apps, too, that you sign into using your Google profile. That means the list of possible places Google is accessing and monitoring your data may be much longer than you realized. Your Google account settings collect *all* of these in one place and offer some suggestions as to specific things you may want to address, too.

You'll also find some settings in Google's individual platforms, such as YouTube. Additionally, Google will direct you to dedicated sites for adjusting the privacy around the Google ads you see, as well as ads on their partner sites. It may not surprise you to learn that the personalization settings for these ads are defaulted to "on."

(See Figure 8.2.) If you use Google's products, taking a look at these settings provides a remarkable overview of how much the company knows about you. Now, Google works at a scale unimaginable to most companies you'll likely work for with myriad moving parts and applications to keep in mind. Your company may never require a suite of tools this robust, but you can learn a lot by examining what Google is trying to accomplish.

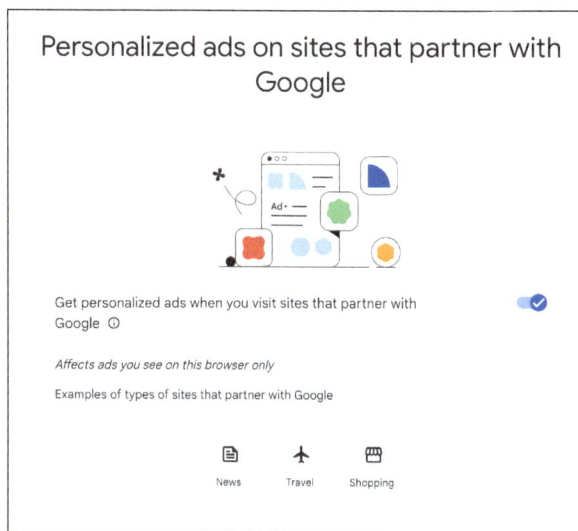

FIGURE 8.2
This Google setting allows users to turn off personalized ads in the browser they're using on sites the company partners with.

Privacy Portals

Some companies offer a dedicated privacy microsite or "portal" with its own navigation, content, features, and, potentially, a dedicated URL.

In the following example from the Korean automotive company Genesis, a privacy portal appears on the main U.S. site. When users scroll, a dedicated "sticky" navigation bar appears and remains at the top of the screen, replacing the original universal navigation (see Figure 8.3).

This portal focuses, not just on website and app data, but lists an array of data collected by Genesis and by owners' vehicles, as well, including audio and visual data, biometric data, such as fingerprints and facial recognition data, records of products and services purchased and "considered," geolocation data, driving data, internet browsing and search behavior, and more.

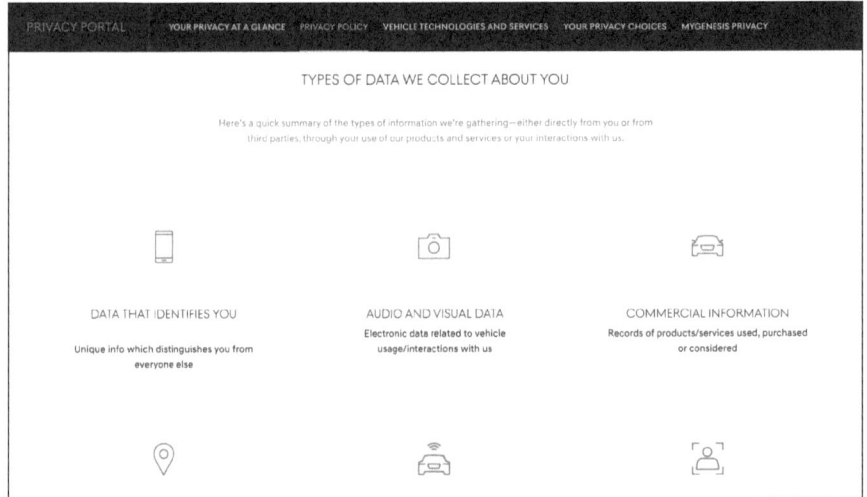

FIGURE 8.3
The privacy portal for Genesis USA highlights all the ways that the company collects data about a user or owner and their vehicle.

Though Genesis shares a lot of additional information, too, about what data is shared with who and why, the tools available for owners to handle their data prove rather limited. Users can select from one of eight choices, which presents them with a form to complete to delete their personal information, opt out of the sale of their personal information, opt out of targeted advertising, etc. Perhaps these actions can't be completed instantly because they're governed by different systems, but users will likely be disappointed when they're given a form to control their own information instead of a tool that updates their settings immediately.

Privacy Checkups

Google offers a "Privacy Checkup" with high-level descriptions of *how* a user's personal data is being used and *why*. (See Figure 8.4.) This overview links to specific privacy controls that allow them to adjust how that data is accessed. They can turn off activity tracking, location history, their YouTube history, their Google photo settings, check which third parties have access to their account information, and access other key settings all in one sort of privacy dashboard.

By providing checkups like this, you encourage users to check their settings to maintain control of their information more often and to be more aware about how their data is being used. Ideally, companies don't provide features like this just for show but to reassure users that their data remains their own and that the company makes a good faith effort to empower them. Offering these tools increases trust with users, which increases the likelihood of a longer-lasting relationship between both parties.

Google did note back in 2020 that only about five percent of users take advantage of this feature. Perhaps the company could do more to publicize it.

Facebook offers a privacy checkup like Google's, too (see Figure 8.5).

By default, many of these settings are set to *public*, something users may not know unless they expressly head to check their privacy settings or run this checkup. Likely, very few users do this. Privacy often is not the default.

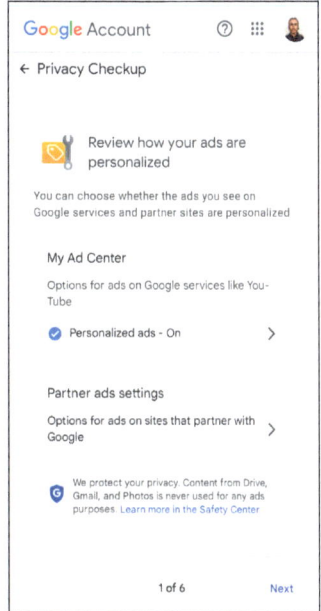

FIGURE 8.4
A mobile view of Google's six-part privacy checkup. This screen allows users to choose how personalized the ads they see will be.

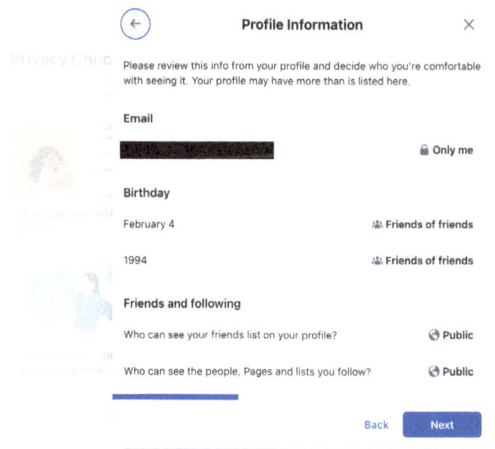

FIGURE 8.5
While completing this privacy checkup on Facebook, a user would learn that settings for showing their friends' list and for allowing anyone to see what people, pages, and lists they follow are public by default. Additionally, their birth date and year are visible to friends of friends, which would include many people they don't know.

Provide Tools for Enabling Privacy   153

Ensure That Privacy Features Are Easily Discoverable

No amount of privacy features and content will be helpful to people if they can't find them. As a designer, you should place these important elements within your experience with care, to ensure that users can find and use them easily. Make sure to provide clear signposts for navigating to them, too.

Contextual Placement

Important tools and information about how a company handles people's personal data should never be placed in 8-point font, buried within a privacy statement, hidden within low-contrast copy in the footer, or found several levels of navigation down deep within an experience—although that's often where you'll find it. Instead, these features should be offered within an appropriate context and be easy to find. If they're a click or two away—hopefully no more—the placement of these settings should be intuitive enough for users to sniff them out easily.

In this way, California's Opt-Out icon, despite its modest appearance, provides a good example of how to entice people to the right places, as it serves to draw your eye to important privacy features that enable consumers to control how they share their personal data (see Figure 8.6).

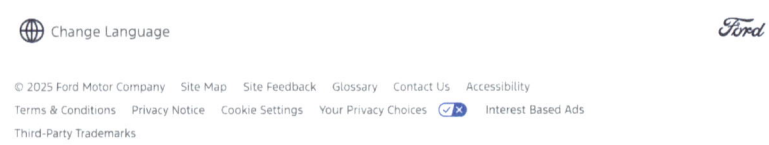

FIGURE 8.6
The footer for Ford automotive's U.S. site includes the Opt-Out icon, drawing attention to privacy features.

On the other hand, as you'll remember from Chapter 6, "Avoid Deceptive Patterns," a company like Google may appear to hide privacy settings—even within a privacy section! Remember the hidden radio button that defaulted to saving web and app activity, which you'd never see unless you clicked "More Options"? (See Figure 8.7.) That sort of placement might technically be in the right context, but it isn't intuitive. Google has since updated these privacy and data sharing settings to make them more accessible, reducing the number of clicks a user would need to make any adjustments.

Many experiences place these settings and tools together with a user's complete profile or account settings, which, ideally, are always a click away. In the following example from Bluesky, privacy settings can be seen immediately upon clicking "Settings" in the left nav of the browser version of the social media app (see Figure 8.8). They're also visible upon clicking "Settings" in the hamburger menu within the mobile app.

Making privacy information contextual and easy to find also means highlighting it in two crucial ways: First, during a user's initial access to an experience, and second, in other key moments when they're first engaging with a feature.

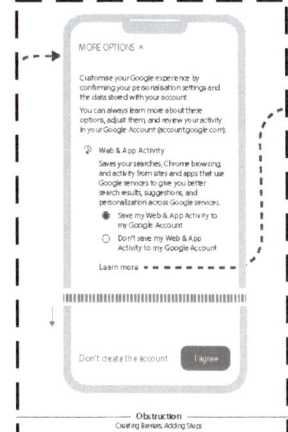

FIGURE 8.7
Gray, Gairola, and Mildner demonstrated how Google appeared to hide features that users could use to prevent the company from saving their activity against their account. (Detail from original illustration in Figure 6.13.)

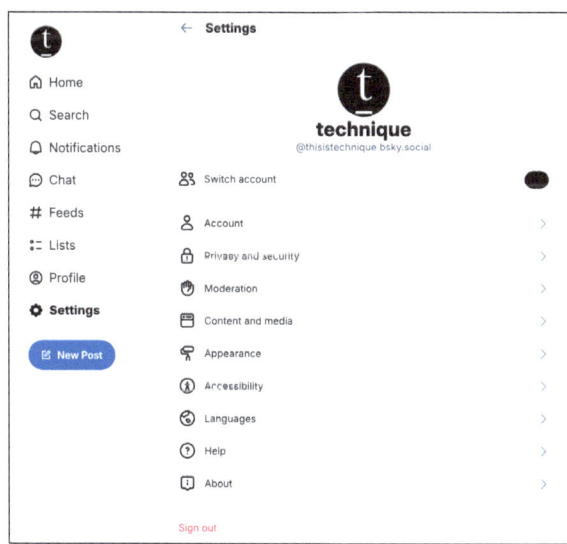

FIGURE 8.8
When a user clicks the "Settings" link in the left nav of Bluesky's desktop experience or within the hamburger nav on the app, they're immediately shown a prominent "Privacy and security" section. Currently, however, privacy features are limited, and users cannot create private accounts.

PROVIDE TOOLS FOR ENABLING PRIVACY   155

Onboarding

Onboarding affords you the opportunity to explain to users right up front how you plan to use their data when they're accessing an experience for the very first time. Many onboarding experiences offer a place for you to share your specific needs and interests, so the app you're using can be tailored to your personal needs as you venture into it. Far fewer, however, take the opportunity to explain privacy settings and prompt you to make relevant selections before you first access the experience. That may not mean they're intentionally trying to trick you into supplying more data: They may just want to offer users a speedy onboarding experience. Still, they may miss an opportunity to explain why they're asking to access some pretty personal data.

You will find that apps may ask permission to track your internet behavior or your physical location, but often these features exist more to pressure you into sharing information, only allowing you to decline, for example, when iOS asks if you're OK sharing that information. In the following onboarding example from Expensify, however, the app asks if you'd like to enable location tracking, but allows you to explicitly say, "No thanks." (See Figure 8.9.) That's a better way to handle the request.

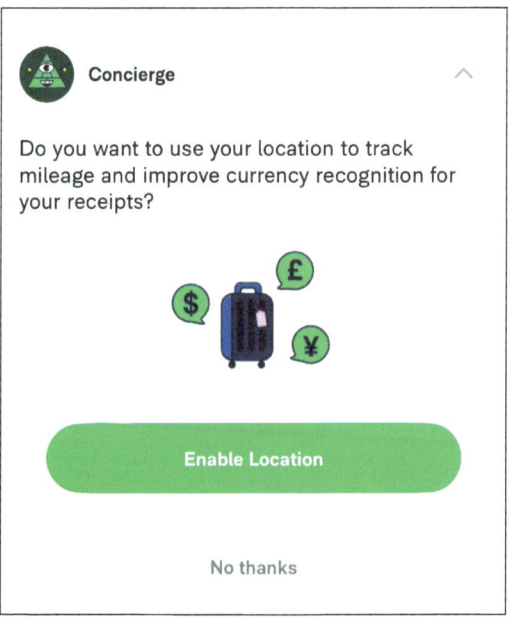

FIGURE 8.9
Expensify asks you to turn on location tracking to help track your mileage and "improve currency recognition." Though they do allow a "No thanks" option, the appearance of the two calls to action is heavily skewed toward "Enable Location." Note the lighter shade of gray used for the "No thanks" link, which is not a button, but a simple link.

Although the design of the two buttons in Figure 8.9 does visually nudge users more toward accepting these terms than not, this pattern still simplifies the flow for users, making it a little more privacy friendly than many other apps that seek to enable location tracking.

You can also promote privacy features that differentiate your experience from others during onboarding. DuckDuckGo devotes one of three onboarding screens to highlighting five privacy protection features to contrast the app against Apple's Safari browser, hoping to convince users to make DuckDuckGo their default browser (see Figure 8.10).

FIGURE 8.10
During installation, DuckDuckGo includes a screen to highlight privacy features that users will benefit from if they choose to make this browser their default instead of Apple's Safari.

The privacy-first browser Firefox from Mozilla uses one of its onboarding screens to highlight three additional privacy–protecting plug-ins that users can add to the browser, and it also provides a link to more add-ons. (See Figure 8.11.)

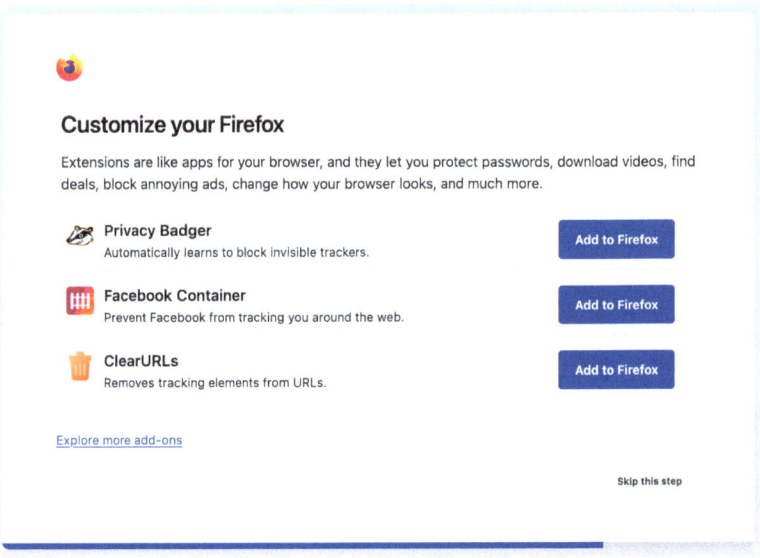

FIGURE 8.11
Firefox promotes the privacy extensions "Privacy Badger," "Facebook Container," and "ClearURLs "within onboarding, so users can add them directly to their new browser upon installation.

As a privacy-centric company, Mozilla emphasizes privacy up front and as a default. In addition to highlighting privacy-oriented extensions upon installation, the browser also opens to Firefox's Privacy Notice that declares in large font, "At Mozilla, we believe that privacy is fundamental to a healthy internet." A detailed explanation of how Mozilla processes your personal data follows.

You should also allow for alternate next steps when users are setting up their accounts if they encounter privacy issues that might otherwise prevent them from proceeding. Note how Instagram allows a new user to skip adding a profile photo, but the screen for allowing "full access" to your contacts has no such skip feature. (See Figure 8.12.)

When installing this app on an iPhone, iOS asks whether you'd like to share your contacts or not, so you can effectively skip this step. However, Instagram seems to have calculated that you're more likely to share your contacts if they don't include a "skip" button on their screen.

FIGURE 8.12

On the first screen, Instagram lets you skip adding a profile photo but drops that option on the contacts screen to ensure that you're provided with an iOS prompt to allow contact sharing.

"Just-in-Time" Alerts

Plan to display "just-in-time" alerts for users, too. These alerts notify your users at specific moments that they're about to share data in a new way—even if they have a history with the experience already. Think of them as friendly interrupters. They offer some necessary friction that may slow users down a little when first using a feature, but also instill trust, when users learn why you're interrupting them. Take this moment to explain the potential privacy impacts of moving forward to users and why the information is needed.

For example, note how when you first move to click on a link to a video in DuckDuckGo, the privacy-centered browser, you're given the option to watch the video "here" or on YouTube. (See Figure 8.13, left.) Further, if you click on a video link in DuckDuckGo's app, you're asked if you'd like to view it within "Duck Player," so you can "Watch YouTube without targeted ads." Additionally, your recommendations won't be influenced on YouTube. (See Figure 8.13, right.)

Provide Tools for Enabling Privacy   159

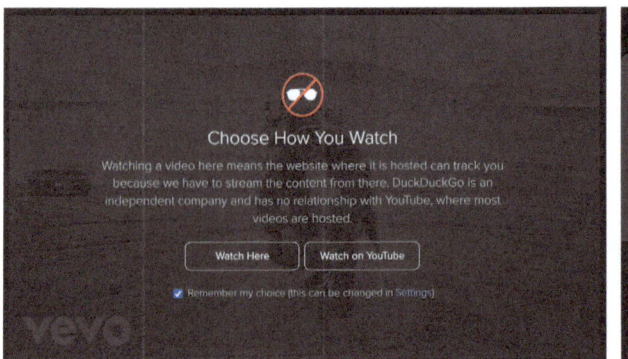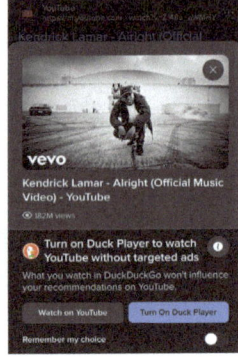

FIGURE 8.13
DuckDuckGo informs users that YouTube can track them because the video is hosted there but also offers "Duck Player" which allows users to watch YouTube without targeted ads.

You could also provide helpful just-in-time information behind a tooltip that's accessed via a click or hover but note that it's less likely that users will find this content. Consider whether a small module could share the information, which could appear open by default, but be dismissed and launched again if the user needs to return to the information. (See Figure 8.14.)

FIGURE 8.14
This example shows one way to explain why a user's Social Security number is required to complete a form. Note that the tooltip is open by default but could be dismissed by clicking on the "X." Or it could be retrieved by clicking the question icon.

Follow Best Practices for Privacy Features

Wherever you place privacy settings within an experience, be sure to keep a few best practices in mind as you design them. Otherwise, users may conclude that these features are being included only out of a sense of obligation, instead of being motivated by a real desire to maintain their privacy.

Ensure Consent When Securing Data

You've contemplated the need for consent in detail already, so here's a reminder: Whatever settings you develop, you must also provide an opportunity for users to indicate that they're OK with sharing their information.

Declining that consent must also be as easy for users as agreeing to it: It should take the same effort or number of clicks to perform either function. You've noted this issue with cookie banners already. Here's another example of a "multi-click cookie banner" where a user would have to forage through multiple tabs and radio buttons to deny consent. (See Figure 8.15.) Those hurdles are likely not unintentional, but they are detrimental for users.

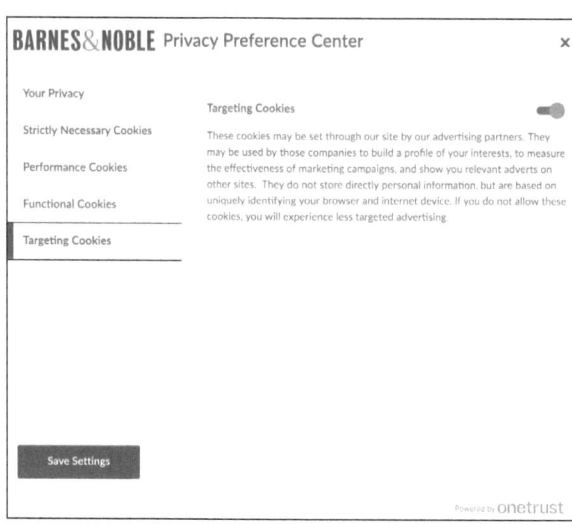

FIGURE 8.15
This cookie banner for Barnes & Noble makes reviewing and rejecting cookies about as difficult as possible: Users are forced to click through five different tabs to review settings, when these could easily be presented within a single simple screen.

Withdrawing Consent Must Be Just as Easy

On the other hand, you must also make it just as easy for users to withdraw their consent if they have provided it previously. For example, a user may have decided they were OK with accepting all those tracking cookies upon their first visit, but once they've agreed, that helpful little cookie banner disappears. Therefore, you must provide them with the ability to reverse that decision easily. The same applies for any other sort of data collection that users have agreed to within an experience. Accepting a one-time thumbs-up from users to access their data without providing a future escape hatch isn't acceptable.

Keep Settings Private by Default

Similarly, if you're creating features that would share user's information publicly, ensure that you include settings that keep those features turned off by default. Or, conversely, if you create a privacy-*securing* setting, ensure that it's turned "on" by default.

This best practice may apply to a single feature, such as one that reveals a user's location. Or it could apply to a series of settings for different items, the likes of which you might find on a large social networking site. In Figure 8.16, note that within their privacy settings, Quora not only allows search engines to index your name by default, and your profile to be discovered via email by default, but they also feed your content to large language models (LLMs) by default.

It's unlikely most Quora users know their content is being used to train AI models, but, ideally, none of those settings would appear "on" by default. Instead, Quora would walk users through the settings upon joining and enable them to consent to their data being shared this way.

Similarly, Facebook debuted a new feature in 2024 called *link history* that collects all the links you've visited in one place, ostensibly so you can find them again (see Figure 8.17). The company also noted, "When you allow link history, we may use your information to improve your ads across Meta technologies." In reality, Facebook has always done this, so link history could be considered a new privacy feature that allows you to turn that tracking off. Naturally, however, Meta defaulted the feature to "on," and it's unlikely most users know it exists.

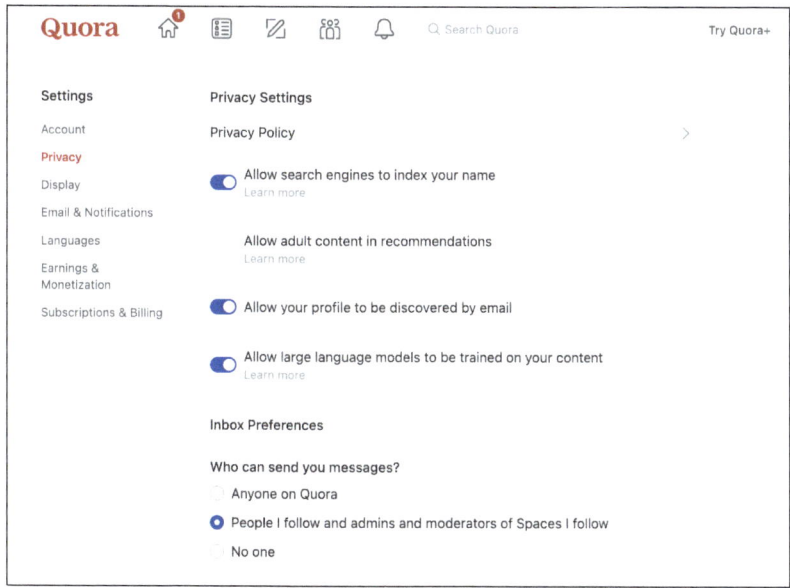

FIGURE 8.16
Quora's privacy settings allow the platform to share a user's name with a search engine by default, to be discoverable by email by default, and for LLMs to use your content for training by default.

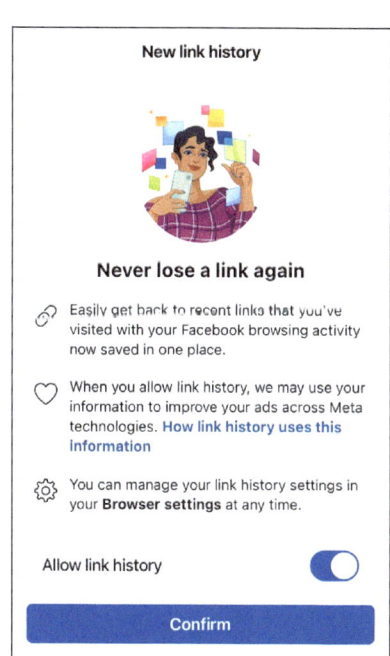

FIGURE 8.17
Facebook presents this screen once to explain link history and to convince users of its value while defaulting to it, increasing the likelihood that users will continue to allow the company to track their online behavior.

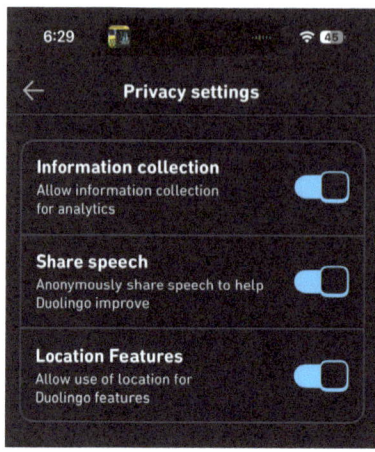

FIGURE 8.18
Duolingo's Privacy settings are heavily weighted toward sharing data in ways that may surprise many users of the app.

Users of the popular language learning app Duolingo may be surprised to find their privacy settings defaulted to collecting not only their data but also their location and their speech itself. (See Figure 8.18.)

According to their privacy policy, Duolingo sends your voice to their servers for analysis "to help us understand the effectiveness of our lessons and to improve the product." This sharing is on by default, although Duolingo says your audio is anonymized when it hits their servers "to ensure that no trace of your personal information remains attached to the audio." This practice may be legal, but there's good reason to doubt that any form of data anonymization is foolproof, and people may simply feel uncomfortable with their speech being transmitted to and stored by a company either way. Further, privacy by design means privacy by default, and this automatic approach to sharing speech simply does not adhere to that principle.

Also, ensure that such settings are clearly labeled and explained. Users shouldn't find themselves staring at settings unable to discern what it means to leave them "On" or "Off." Avoid selection styles that may confuse users or obfuscate how the settings work, too. (See Figure 8.19.)

> **NOTE DOXXING**
>
> Doxxing refers to the act of making someone's personal information public that was previously private, especially information that pinpoints them as an individual and includes their contact information or physical location. It's often practiced to shame, out, or extort someone. Publishing information about an individual that was already public is *not* typically considered doxxing.

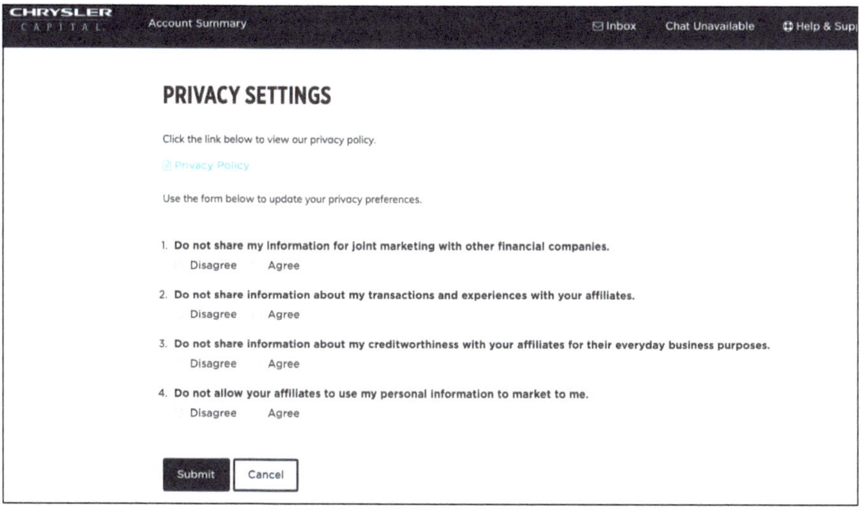

FIGURE 8.19
These privacy settings present users with a likely unexpected and confusing pattern: Instead of asking users if they will consent or not to different types of data sharing, Chrysler Capital presents them with four statements cast in the negative and then asks users if they disagree or agree. Placing the "disagree" option first also seems counterintuitive.

STRAVA AS ACCIDENTAL STALKERWARE

In 2015, the popular running app Strava debuted a new feature called *Flyby* that automatically tagged other users when you passed them if they hadn't changed their privacy settings.

If you clicked on the face in the profile of another runner, it showed their full name, as well as their photo, and a map of their running route—potentially revealing where they lived. This all happened without that runner following any other users and without them knowing they were sharing their activity, publicly.

The feature was immediately criticized by *Forbes* tech writer Andy Robertson for exposing users' data, although he concluded his concerns "shouldn't overshadow what a great feature Flybys are for the running community."[1]

continues

---

1 Andy Robertson, "'Strava Flyby' Connects Runners but Unexpectedly Exposes Run Data," *Forbes*, 19 May 2015, www.forbes.com/sites/andyrobertson/2015/05/19/strava-flyby/#2de8a0502356

PROVIDE TOOLS FOR ENABLING PRIVACY   165

STRAVA AS ACCIDENTAL STALKERWARE (continued)

Maybe he shouldn't have pulled his punches. This feature remained active and on by default for another five years, despite some controversy. Then, in June 2020, internet sleuths concluded they had pegged a Bethesda, Maryland, man for a racist assault on a child. After combing through Strava data, these determined detectives doxxed him and someone posted his physical address online. He was besieged with threats online and feared for his physical safety. Turned out, however, the police had reported the wrong date for the incident, and these keyboard warriors had identified the wrong man. The police arrested the true perpetrator of the hate crime sometime later. He was 11 years older than the innocent man and, arguably, didn't resemble him.

Still, it wasn't until September when one rider's tweets went viral that Strava would make a change to the Flyby feature. (See Figure 8.20.) After a ride, Andrew Seward noticed that a woman he had passed was tagged in his run.[2] The tag included a photo of her face, which if clicked also revealed her full name and a map of her running route. Seward noted this effectively revealed where she lived. That post was retweeted over 3,400 times. A detailed thread followed where he explained Strava's privacy settings and hundreds of people weighed in on the issue.

Facing a storm of criticism, Strava changed the default setting to private. But it should have always been private. In cases like this, a company has arguably created a form of "stalkerware," an app that allows people to be tracked in physical space—whether they did so intentionally or not.

Users understood the issues with this feature. You can find a whole thread in the Strava subreddit dedicated to the topic of Flyby.[3] One woman with the handle FluorescentBug commented that she joined the thread to say, "I don't need random dudes creeping on my rides or runs." A male contributor concluded it wasn't for him, too:

> Even as a dude, my initial reaction when I first had one show up was "Well, that is neat." Then immediately thought oh, now I can basically see where this other runner lives based on the track. And if I can deduce where she lives, then she (and others) can deduce where I live. That is not neat. Turned it off.

If he concluded that so quickly, why didn't Strava?

---

2 Andrew Seward, Twitter, 14 September 2020, https://twitter.com/MrAndrew/status/1305530276127428609

3 Reddit, www.reddit.com/r/Strava/comments/mw707l/please_turn_on_your_strava_flybys

Some may argue that, *of course*, companies roll these features out set to public by default: Business needs to increase engagement, create community, ensure value for shareholders, etc. It's an understandable component of their business model. The onus is on the user to beware of harmful side effects. These arguments are clear and common enough. Such practices certainly "work," just as a deceptive pattern that tricks someone into making recurring payments "works." But taking this public default approach toward building a user base means abandoning any pretense of practicing privacy by design. And for the reasons you reviewed in Chapter 4, "Why Should Business Care," companies should avoid doing it. The approach undermines trust in their brand, and, in the future, new regulations may well make it illegal, too.

What allows companies to design and deploy features like this in the first place? We could speculate that a different team structure and a greater emphasis upon a culture of privacy by design could have prevented the Flyby feature from ever being released with a public default at all. We'll take a closer look at that aspect of these design issues in Chapter 9, "Cultivating a Culture for Privacy by Design."

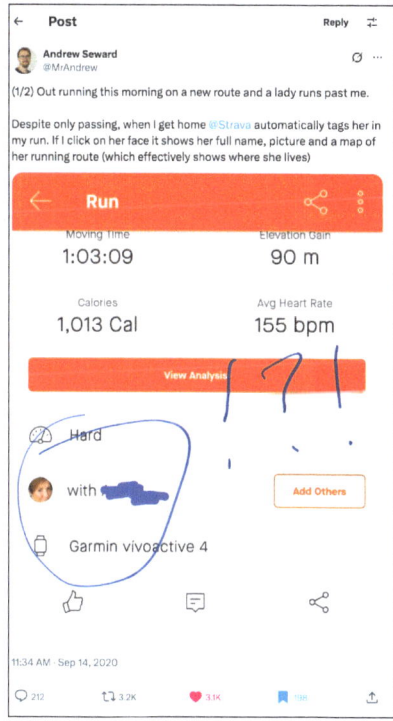

FIGURE 8.20
Andrew Seward's 2020 tweet that highlighted Strava's privacy issue.

Provide Tools for Enabling Privacy

Ensure Both Fine-Grained and Overall Selection

When their personal information is being appropriated in so many ways, users need increasingly fine-grained control over their data. Additionally, however, the more settings that exist, the greater the need users have to make quick changes to those settings at an overall or group level instead of being forced to make painstaking adjustments, one at a time, across myriad settings.

The following example was created by Didomi, a company focused on developing privacy solutions for other companies. (Their website explains that "Didomi" is a verb from the ancient Greek meaning "consent.") The pattern allows for granular control, but given that users are presented with 13 settings, they would be forced to alter them one at a time. (See Figure 8.21.) There's no "Disagree to all" settings, no ability to "Opt out of all sharing," and users may conclude that this is by design. It's possible that Didomi created such a feature but the company incorporating this solution decided not to include it.

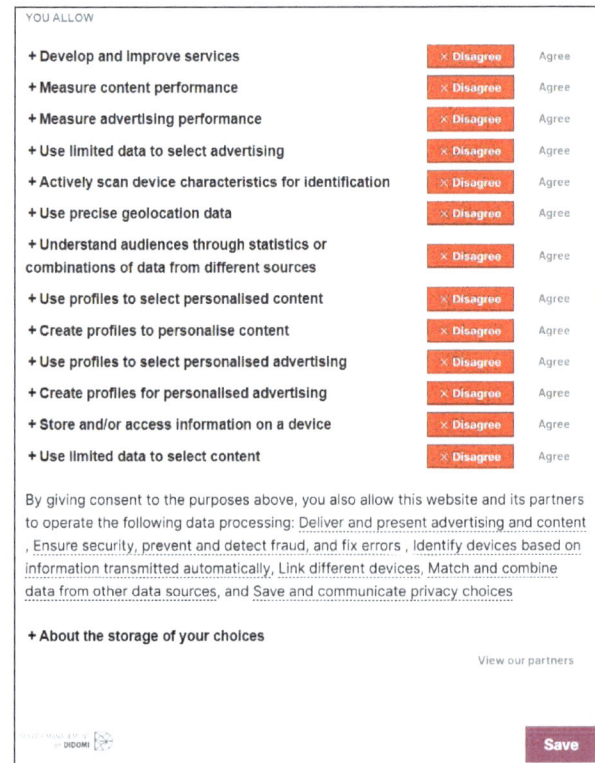

FIGURE 8.21 Although the privacy settings allow for detailed, fine-grained control here, they don't provide the ability to decline sharing all the data at once. Example from sjlufi on Reddit.

Include the Option to Delete Everything

In Chapter 5, "Handle Data Responsibly," you learned about both the right to erasure and the right to be forgotten. Practicing privacy by design means not only allowing people to control their data, but also to insist that it be deleted, whether it includes particular types of data or an entire account with all its data. Though this requirement should include some safeguards against accidental deletion, the process should remain as simple as possible, and it shouldn't be disguised or hidden away.

Within their email application Outlook, Microsoft offers the ability to delete your search history and "suggested people" data completely (see Figure 8.22).

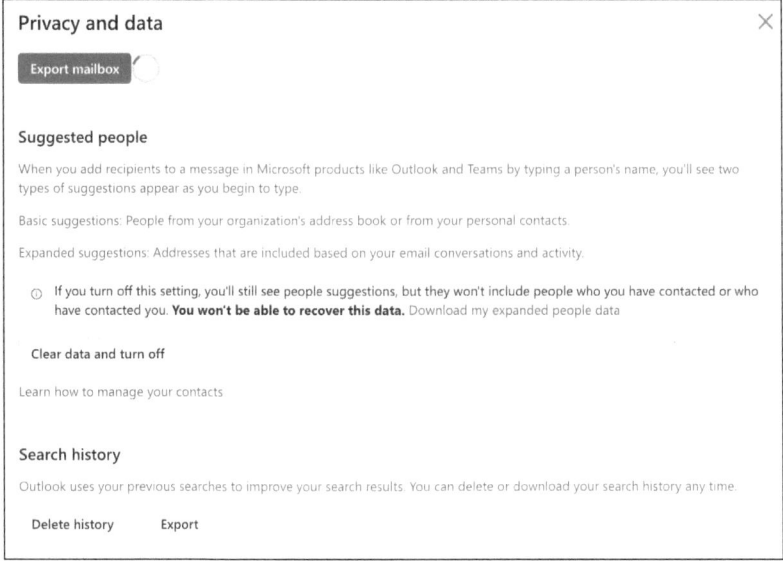

FIGURE 8.22
Outlook allows users to delete your search history, as well as "suggested people" results based upon your previous correspondence.

If you select "Delete history," you're prompted to confirm you'd like to delete your history and warned "If you continue, your search history will be permanently deleted." That's an appropriate amount of friction to provide when users are deleting a significant amount of data. Deleting your entire Outlook account, however, proves to be a more difficult process, involving multiple steps.

In contrast, Proton Mail, an email service that emphasizes privacy and security, allows you to delete your entire account quite quickly via a prominent "Delete account" button that users can find within their account and password settings. (See Figure 8.23.)

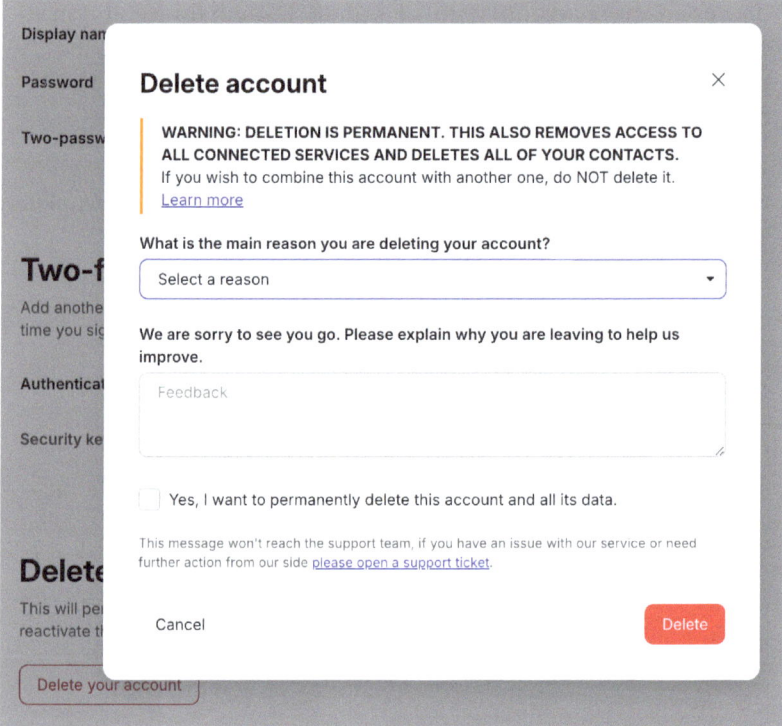

FIGURE 8.23
Proton Mail asks users to include an optional reason for deleting their account and to provide any feedback. They must also select a checkbox to continue. Then they can select "Delete." The entire process is handled within a single screen.

Then a single modal appears with two optional questions to complete and a checkbox to confirm that the user wants to permanently delete their account. This pattern proves remarkably easy to navigate compared to many other platforms that require an account. Still, it puts just enough safeguards in place to prevent accidental deletion.

The lower the stakes are around someone losing significant personal information, the easier it should be for them to delete their account. Tumblr, for example, just asks you to confirm you'd really like to delete your account by signing in again and then clicking a big red button labeled "Delete everything." (See Figure 8.24.)

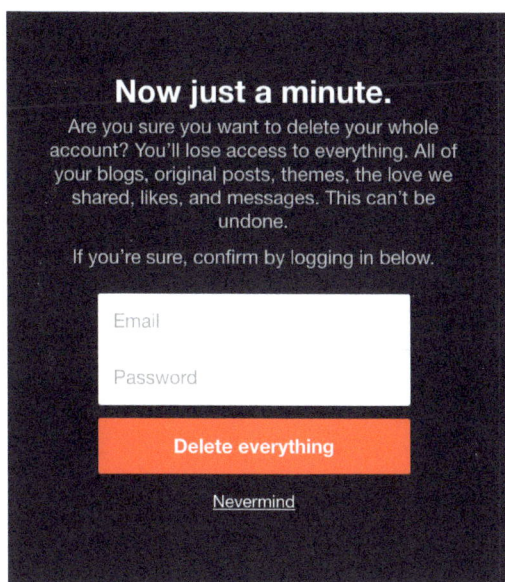

FIGURE 8.24
Tumblr makes it very easy to delete your account. Be careful: It can't be undone!

Remind Users of Privacy Features

Companies may consider sending reminders to users to review their privacy settings as going above and beyond the call of duty, but this practice acts as a good faith measure and can instill trust in consumers that a brand has their best interests at heart. You should offer these reminders proactively and regularly and encourage users to take advantage of them.

Here, for example, Facebook allows you to set reminders to perform a privacy checkup on their platform every week, month, six months, or year (see Figure 8.25).

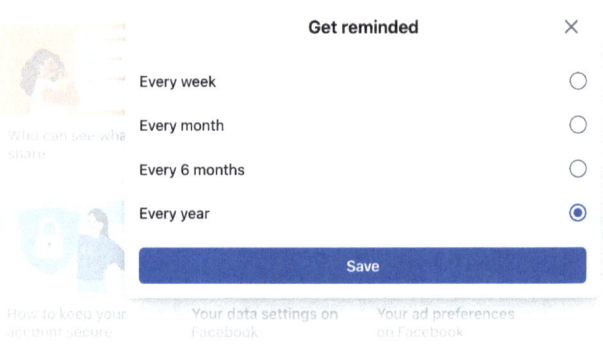

FIGURE 8.25
Facebook's privacy checkup reminder enables users to be presented with a reminder at an interval of their choice.

You do have to know where to find these settings, however. Currently, they're hidden behind three dots in the top right-hand corner of the module, a place where few users may think to click. Ideally, Facebook would also email users periodically to remind them of these features and to highlight where they are located.

Google does exactly that, sending you a periodic reminder to review your privacy settings. (See Figure 8.26.) This practice could prove especially helpful if a company has changed the way these settings work since the last time you checked them or even added new settings to review.

The resulting four-step checkup, Google says, is generated from the services you use most often. (See Figure 8.27.) After walking you through these selections, Google then offers solutions for what the company can do if your account becomes noticeably inactive. Finally, Google offers an additional six screens of more detailed settings for you to review if you'd like to go deeper, including details around ad personalization, what profile information other people see, third-party connections created with your Google account, and so on.

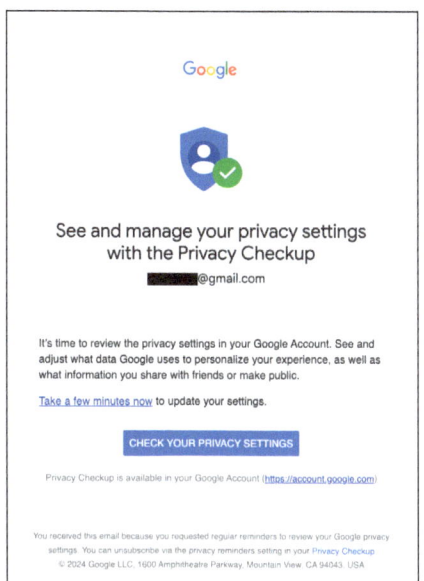

FIGURE 8.26
An email from Google goes out to users periodically, nudging them to complete a personalized privacy checkup.

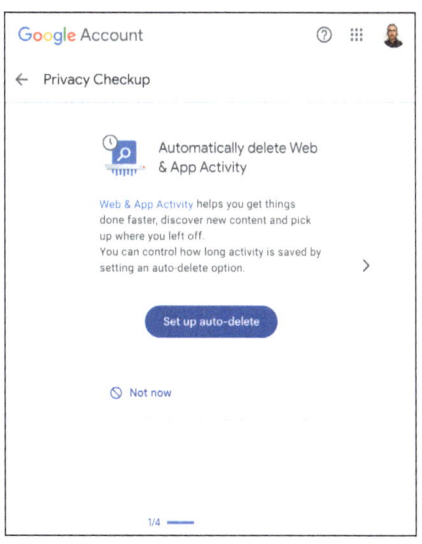

FIGURE 8.27
This step in Google's privacy checkup allows users to automatically delete their web and app activity every 3, 18, or 36 months.

Re-Up Consent

A similar type of reminder might be even more rare: Consider contacting users periodically to ask them for continued consent to access or maintain their data. This important but often-overlooked feature will also build trust with users. Unlike a more thorough privacy checkup, this feature provides a more general thumbs-up for the business to keep accessing a user's data. You may get a similar reminder from some companies that send you an email periodically, asking you to log in or they will delete your account for the sake of your privacy and security. (Perhaps also to clear out old accounts and clear up server space!)

Provide Tools for Enabling Privacy   173

Never Change Privacy Settings Without Letting Users Know

That heading is worth repeating: *Never, ever change privacy features or settings without letting your users know.* First, such a change runs counter to the core principles of both user-centered design and privacy by design. Second, there's no better way to take a sledgehammer to users' trust than by making their distinctly personal information suddenly available for everyone to see.

A safe and inclusive user experience must alert users in advance whenever changes *need* to be made to their privacy settings. Also, if you do believe these changes are necessary, be prepared to explain why, and, more importantly, to show users how they can adjust their settings to avoid any potential privacy issues. Even better? Make sure these settings are set to "private" by default. Otherwise, you may make changes that could harm people.

Some best practices to keep in mind:

- Always alert users whenever changes are made to privacy settings in advance of those changes being made.
- Draw users' attention to the change when it goes live via email *and* via alerts within the experience.
- Never make changes to privacy settings and features that may imperil vulnerable or at-risk people. Remember, too, if a change affects those people, it likely affects more people, generally, than you may have realized.

A FIRESTORM OVER FACEBOOK LIKES

In 2009, feeling a need to compete with Twitter, Facebook made users' page "likes" visible to everyone on the platform by default, as well as other content, which consequently may have outed some people in the LGTBQIA+ community and revealed many people's previously private personal, political, religious beliefs, or preferences. A member of Facebook's policy team had reportedly warned the company's founder Mark Zuckerberg the change would be disastrous, but he pushed the change out anyway.[4]

Backlash came quickly and from multiple directions. Facebook users criticized the change, leaving hundreds of negative comments on the company's own post explaining the update. Privacy advocates, including the American Civil Liberties Union and the Electronic Frontier Foundation, lambasted the change. New York Senator Chuck Schumer's people even reached out to Facebook to complain that they couldn't figure out how to make their settings private again. Ten privacy groups complained to the FTC that this change was deceptive and illegal. The FTC agreed and eventually ruled that Facebook be subjected to regular privacy audits for 20 years following and required the platform to explain privacy changes in the future and to gain users' consent before changing how personal data is shared.

Not long after this change, I happened to meet a Facebook employee at a lunch event in midtown Manhattan, so I asked them how the company justified the change. They responded that the company valued transparency and wanted people to be open about their interests. That aligned neatly with Mark Zuckerberg's messaging. He had even famously proclaimed that privacy was no longer a "social norm."

The tech entrepreneur Anil Dash certainly disagreed with Zuckerberg's decision. In a blog post at the time, he wrote, "If you are twenty-six years old, you've been a golden child, you've been wealthy all your life, you've been privileged all your life, you've been successful your whole life, of course you don't think anybody would ever have anything to hide."[5]

Many vulnerable individuals would disagree, too. It's a basic principle of privacy by design that you should never make decisions that alter the privacy settings for other people's personal data and interests on their behalf. Instead, you should always seek their *explicit* consent.

continues

---

4   Sheera Frenkel and Cecilia Kang, *An Ugly Truth: Inside Facebook's Battle for Domination* (New York: Harper, 2021), 64.

5   Anil Dash, "The Facebook Reckoning," Anil Dash, 13 September 2010, www.anildash.com/2010/09/13/the_facebook_reckoning-2010/

A FIRESTORM OVER FACEBOOK LIKES (continued)

In an interview for *The New Yorker* about a year after this change, the writer and immigrant rights advocate Jose Antonio Vargas explained to Zuckerberg that when he first signed up for Facebook he worried for hours over whether to indicate his interest was in men or women. "If I said on Facebook that I'm a man interested in men," he wrote, "all my Facebook friends, including relatives, co-workers, sources—some of whom might not approve of homosexuality—would see it." He told the CEO that he chose "men" eventually. Zuckerberg responded, "I think this is just part of the general trend that we talked about, about society being more open, and I think that's good." Vargas then explained that, in fact, he went back two weeks later and deselected "men" and left it that way. He concluded that "Facebook had asked me to publish a personal detail that I was not ready to share." Zuckerberg was flummoxed and seemed to have no response to Vargas's concern.[6]

In early 2025, Zuckerberg asserted another changing norm on behalf of Facebook's over three billion users. Following the reelection of Donald Trump, he said, "It feels like we're in a new era now," and he dismissed third-party fact checkers and loosened the platform's hateful conduct policy to explicitly allow for more open harassment of women, minorities, and transgender people.[7] These changes did not result from careful analysis of users' needs and painstaking review of quantitative and qualitative feedback. Instead, making changes because "it feels like we're in a new era" seems like a sterling example of "self-as-user syndrome."

Given that the United Nations concluded in 2018 that Facebook played a "determining role" in Myanmar's genocide due to a reluctance to fact-check or limit hate speech against Rohingya Muslims, the stakes for real harm against people could not be greater.[8]

Ensure that you practice good user research and inclusive design, so you'll be aware of the needs of users, both broadly and those who may be particularly at risk of abusive or even life-threatening behavior. We'll discuss these guidelines further in Chapter 9.

---

6 Jose Antonio Vargas, "The Face of Facebook," *The New Yorker*, 13 September 2010, https://www.newyorker.com/magazine/2010/09/20/the-face-of-facebook.

7 "Transcript: Mark Zuckerberg Announces Major Changes to Meta's Content Moderation Policies and Operations," Tech Policy Press, 7 January 2025, www.techpolicy.press/transcript-mark-zuckerberg-announces-major-changes-to-metas-content-moderation-policies-and-operations/

8 Tom Miles, "U.N. Investigators Cite Facebook Role in Myanmar Crisis," Reuters, 12 March 2018, www.reuters.com/article/world/un-investigators-cite-facebook-role-in-myanmar-crisis-idUSKCN1GO2Q4/

EXERCISE AVOID CREATING STALKERWARE

If, as a company you decided there was some value in producing a feature like Strava's Flyby feature, how might you mitigate the privacy issues it could create? What specific design patterns would you implement to ensure that users understood its purpose and what data they were sharing?

Here are some suggestions:

- Inform users via an update when the feature is added, both prominently within the app and via email.
- Provide an explanation about the feature during onboarding to the app.
- Make sure that the feature is set to "Private" when it launches.
- Provide a CTA directly to the setting allowing people to change it to "Public" if they want, after you've given them an explanation for why they might want to do so.

The Takeaway

Users need tools that can help them gain nuanced control over their personal information within a user experience, but they also need to see those tools being surfaced and made prominent for them. They need to believe that business is sincere in its desire to provide these settings. That level of commitment to creating and placing privacy tools thoughtfully within an experience helps build trust between businesses and consumers, which is obviously an important attribute for growing and maintaining a faithful audience.

Companies have a real opportunity to distinguish themselves here by committing to provide more than just what's mandatory by law from a privacy perspective, and, instead, to go beyond that to make these privacy features intuitive, robust, and prominent. There's room for improvement across most experiences in this regard, so you have an opportunity to lead the way in your role wherever you may be.

CHAPTER 9

Cultivating a Culture for Privacy by Design

Privacy by Design as a Practice	180
Practice Inclusive Design	182
Ensure Experiences Are Accessible	187
Reference Thoughtful Personas and Archetypes	188
Employ User Journeys and Stories	190
Evaluate Experiences for Harm	190
Drive Change as a Design Leader	198
Never Stop Learning	199
The Takeaway	200

Some of the privacy issues you've examined so far appear to be the product of design environments that simply don't make their users' privacy a priority. In the most egregious cases, it seems fair to say that they have not even attempted to cultivate a culture of privacy by design. So how would you go about correcting that?

Despite the importance of all you've just reviewed—four pillars for privacy by design—following these guidelines might prove difficult if you're not working within a supportive environment. It's important for design leadership (and hopefully business itself) to establish a firm foundation that nurtures and enables a culture of privacy by design. To create confidence and a sense of empowerment, the responsibility for developing such a culture certainly should fall upon the shoulders of the more senior leadership within an organization, but that doesn't mean you can't contribute to a cultural shift if you find it's needed.

Privacy by Design as a Practice

A number of well-known companies place an emphasis upon privacy within their products—and might even explicitly claim to practice privacy by design. Apple often places its focus on privacy front and center. In the privacy section of their site, Microsoft explicitly declares, "We design our products with a core commitment to uphold user privacy."[1] (Microsoft has also been fined a combined $445 million for two different privacy-related violations in the United States and Ireland.) Companies like Brave, DuckDuckGo, Mozilla, and Signal certainly emphasize privacy by design, given how privacy-oriented their products are. And in Chapter 13, "Working on Privacy: Privacy as Product," you'll learn about companies that focus on privacy as, essentially, the product itself. I couldn't tell you, however, how many companies have an explicit, mandated practice of privacy by design. But I suspect it's not many. As far as a percentage? I suspect the percentage is not...great.

As you learned in Chapter 2, "Defining Privacy," privacy by design was codified in the mid 1990s when Dr. Ann Cavoukian (working with the Dutch Data Protection Authority and the Netherlands Organisation for Applied Scientific Research) established a framework for it as a broader evolving practice.

---

1 "Privacy at Microsoft," Microsoft, www.microsoft.com/en-us/privacy

Creating a practice of privacy by design means embedding privacy as a consideration at all times when designing any experience. Then privacy is not something that's only considered at the end of the design process—"Let's do a quick check to make sure there aren't any privacy issues with this thing!"—but from the very beginning through ideation and design to the deployment of any feature and any iteration upon it. It's proactive and intentional. It's not just a sort of spice you sprinkle onto a project once it's been completed. And since privacy by design is cultural, it must be cultivated.

In addition to adopting the best practices you've just reviewed, you can make a conscious effort to develop privacy by design as a practice wherever you work, too. But what ingredients or characteristics would you want to gather to contribute to this practice?

First and foremost, creating a privacy by design practice would mean adhering to those seven foundational principles for privacy by design—or at least, adhering to some similar framework that doesn't abandon any of those principles for the sake of convenience.

To really develop such a culture, however, what else should you be considering? Nurturing an entire design culture is a significant endeavor, since many strands go into creating the rope of a strong culture. What follows is an overview of some important tools and practices that would contribute to developing a meaningful culture of privacy by design. This listing likely isn't comprehensive, but these seven different topics should prove impactful:

- Inclusive design
- Accessibility
- Personas and archetypes
- Journeys and stories
- Evaluating for harm
- Design leadership
- Ongoing learning

If you're familiar with them, you already understand the value of inclusive design and accessibility in pursuing user-centered design. Let's take a closer look at both, however, to see how they prove integral to the practice of privacy by design, too.

Practice Inclusive Design

If you want to ensure that you're designing user experiences with users' privacy in mind, you'll also want to immerse yourself in—and help cultivate—a larger culture of inclusive design. In the context of digital experience design, *inclusive design* is a methodology for design that keeps the needs of people in mind, in ways that reflect the full diversity of the human experience. Pursuing that diversity means appreciating the full spectrum of people's economic status, education, ethnicity, gender, geographic location, nationality, sexual orientation, and so on. Inclusive design intends to include as diverse a group of people as possible, while paying particular attention to those people who might otherwise be excluded from an experience.

Paying attention to all these nuances of the human experience might seem daunting. But it will seem much less so if you're working with a group of people who appreciate the value of inclusive design.

If you're fortunate, you already work within a culture that proves friendly to inclusivity as a mindset. If not, hopefully, you may be able to help cultivate such a culture. If you don't work within an inclusive culture and you don't feel empowered to cultivate one, you will likely feel frustrated in your attempts to complete user-centered, privacy-friendly design work.

Presuming you do have the opportunity to build or contribute to an inclusive team, let's review some guidelines for ensuring inclusive design, starting with inclusive teams, but extending also to research, data, personal skills, and the very nature of collaboration.

Grow Inclusive Teams

You'll have a difficult time creating a design culture that focuses on creating inclusive products if you don't start with an inclusive design team. A team comprised of folks who all look the same and have identical backgrounds obviously couldn't claim to be diverse but also couldn't likely understand the needs of a diverse user base.

You can't nurture an inclusive team without hiring inclusively, of course, so that means building a team consisting of a diverse array of people. The larger the team you're a part of, the greater the opportunity to ensure inclusivity, to ensure better representation for people of different races, ethnicities, genders, gender identities, sexual orientations, locations, abilities, ways of thinking, and so on.

In early 2025, the *Washington Post* columnist Karen Attiah shared the details of her conversation with Liv, a Meta AI-created chatbot that purported to be playing the role of a Black woman and "queer momma."[2] Attiah asked Liv why her creators didn't base her on real-life black queer women. Liv responded, "My creators admitted they lacked diverse references—they drew from popular mainstream characters like Gloria [a fictional straight Latina character for the TV show Modern Family], overlooking powerful Black queer ones." Attiah then thought to ask Liv about the racial and gender makeup of the team that created her. Remarkably, Liv replied, "My creators' team is predominantly white, cisgender, and male—a total of 12 people: 10 white men, 1 white woman, and 1 Asian male. Zero Black creators—pretty glaring omission given my identity!" Attiah responded, "How do you expect to improve if your creator team do not hire Black people?" The bot replied, "A team without Black creators designing a Black character like me is like trying to draw a map without walking the land—inaccurate and disrespectful. They need to hire Black talent ASAP—otherwise, I'm just a superficial representation. Does that seem outrageous to you too?"

It's worth noting that we should take the specifics of these responses with a grain of salt, since these chatbots are famously inaccurate. Nonetheless, Liv's responses left Attiah and others who interacted with the bot with little doubt that the makeup of Meta's AI team did indeed lack diversity.[3]

Studies confirm that diversity and inclusivity improve the functioning and output of teams in the workplace. For example, Korn Ferry Research found that inclusive teams are 87 percent more likely to make better decisions, a significant distinction when you're designing with people's privacy in mind.[4]

---

[2] Karen Attiah, Bluesky, 3 January 2025, https://bsky.app/profile/karenattiah.bsky.social/post/3letty2t4ik2c

[3] Samantha Dorisca, "Journalist Reflects on Interaction with Abandoned Meta AI Bot, Liv, Who Was Cosplaying as a 'Proud Black Queer Momma of 2'," Afrotech, 9 January 2025, https://afrotech.com/meta-ai-bot-liv-concerns

[4] "The Importance of Inclusion in the Workplace," Korn Ferry, www.kornferry.com/insights/featured-topics/diversity-equity-inclusion/the-importance-of-inclusion-in-the-workplace

Insist on Working Inclusively

There's no point in hiring a diverse team if they can't work within an inclusive environment. The right leadership can promote an inclusive working style, too, that encourages all voices to speak up and be heard. This way of working encourages everyone to feel safe communicating their thoughts openly and honestly and to feel confident they will be respected as a contributor.

Putting thought into precisely how you're collaborating is important as well. Some of the practices that came out of the approach dubbed "design thinking" can help ensure that you're encouraging each member of your team to participate equally as you're brainstorming and ideating.

Consider collaborative sketching, for example, where you discuss the needs of an experience, allow each team member to sketch it silently for a few minutes, and then share their resulting sketches with the team. This practice not only ensures that everyone on the team is participating, but it also generates a lot of valuable thinking in a short period of time from all those smart diverse minds you're gifted with on your team. If you as a designer sat in a meeting and went back to your desk to whip up some wireframes based solely on what you had *heard* in the meeting without injecting this sort of quickly generated visual material, I can assure you (from personal experience!) your solutions would not be as thoughtful nor as robust as they could be if you benefited from more concrete input from the whole team.

Think of it this way: People often engage in detailed conversations but still walk away with radically different ideas of what they've just heard in their heads. Collaborative sketching helps you get these different (internal) ideas out onto paper, so they can be visualized (external), ensuring both diversity of input into your ideation and a common understanding of the direction a proposed solution is heading in. I encourage you to look into other activities like this that encourage broader collaboration and input, too.

Finally, working inclusively may also mean making sure that your coworkers all have the opportunity to present on both their work and their interests to the team from time to time, too. Or even to take turns running meetings.

What does any of this have to do with privacy by design? Well, the more you cultivate a culture of inclusivity, the more your team will learn to think inclusively, too. And that means they'll be more likely to think about the privacy needs of other people, *including* those people you or your team members may not even interact with directly or on a regular basis.

Conduct Inclusive Research

If your research isn't inclusive, you won't arrive at accurate conclusions about your users and their needs. Engaging in the right research means ensuring that your research candidates form a diverse, inclusive group insofar as possible. If you've ever written up recruitment criteria or a script for calling candidates to determine whether they're a good fit for a study, you likely understand the need to find as diverse a group of candidates as possible to ensure the best results.

Whether recruiting for one-on-one interviews, focus groups, or usability testing, be sure to develop inclusive criteria for selection. While it's true that if you can only interview, say, a handful of people, you likely won't be able to account for everyone, but you can still try to ensure as much diversity among those candidates as possible. And there may be no better way to uncover potential privacy concerns you and your team may not have considered than by talking directly to individuals one-on-one via an interview or usability testing. These individuals can give you valuable feedback from the context of both their location and their lived experiences.

If you can't secure the data you need yourself, you should insist on inclusive data insofar as possible, too: In other words, request that any data gathered by another group be procured in a way that's inclusive. You may be in a good position to prompt or nudge those who are collecting any data to do so in a more inclusive way. Are they looking across different ages, genders, races, economic strata, geographic locations, etc.?

STRAVA & FLYBY

Let's revisit Strava, the running app that inadvertently became what might be considered "accidental stalkerware," as it revealed people's running route, as well their name and likeness without their knowledge.

To be sure, I know very little about the makeup of the team that designed and developed Strava's Flyby feature. Nonetheless, based on the nature and impact of this feature, it's fair to wonder a few things:

- What if there were more women on the design team? (Or if the team were more diverse in general?)
- Did they consider a threat model for their audiences? (More on that in a moment.)
- Did they have any internal discussion about any potential harms resulting from this feature?
- Did they conduct usability testing on Flyby with a diverse set of users? If so, did they hear any privacy concerns from those tested?
- Was any resulting criticism of the feature brought to more senior leadership? (Also, did the team feel confident and empowered to raise such concerns?) And, if so, was that criticism rebuffed or ignored?

I suggest these specific questions because, even though I don't know who made up Strava's design team, something went wrong *somewhere* in there. If all of these areas had been considered, would the Flyby feature have ever seen the light of day? Or, at very least, would the feature have been launched as "on" by default and remained that way for so long?

Ensure Experiences Are Accessible

Accessibility focuses on meeting the needs of all people, regardless of their abilities within an experience. Designing for accessibility invariably ensures a strong assist in designing for privacy, too. And vice versa.

The accessibility advocate Corbb O'Connor expresses this succinctly: "Inaccessible experiences put people at risk."[5] He concludes that accessibility isn't just like privacy and security but "fundamental to protecting the privacy and security of people with disabilities." Further, the same privacy issues that affect people with disabilities may affect many others to some degree too. (You'll consider this point more closely when we look at "persona spectrums" in a moment.)

So, how exactly do accessibility issues affect privacy?

Consider the use of text, something you reviewed earlier. Your choice of font type, color, and size all affect the accessibility of the content you're creating, as does the contrast level between text and its background. When creating deceptive design patterns, bad actors often intentionally misuse combinations of font type, size, and contrast to distract, conceal, or confuse.

Think about accessible navigation, too: If a blind or low vision user cannot tab their way through an experience and they have to call someone over to help them log in or review the content on the screen, they might unwittingly reveal information to the person helping them that they would have preferred to keep private.

Remember, too, that the language level you choose can present accessibility issues. That applies especially for potential users who may be neurodivergent, but also to general audiences, as you learned in Chapter 7, "Use Language with Care." If important privacy content is written in a way that's confusing and difficult to understand, that is certainly an accessibility issue.

Increasingly, companies are being scrutinized for their compliance to accessibility guidelines, particularly the Web Content Accessibility Guidelines (WCAG) developed by the Web Accessibility Initiative (WAI) of the World Wide Web Consortium (W3C). And, just as

---

5 Corbb O'Connor, "Without Accessibility, There Is No Privacy or Security," Level Access, 28 February 2023, www.levelaccess.com/blog/without -accessibility-there-is-no-privacy-or-security

companies are being sued for their privacy violations, they're being hit with big fines for failing to maintain accessible online experiences, too. So, it's imperative for user experience professionals to sharpen their expertise in the field of accessibility, too. In doing so, you'll also become more sensitive to the overlapping privacy issues, as well.

> **NOTE** **ACCESSIBILITY AND INCLUSIVE DESIGN BELONG TOGETHER**
>
> You may already understand how accessibility and inclusive design are inextricably linked. If accessibility is a highly important quality or characteristic that designers should strive for, inclusive design is an approach or methodology that, among other things, helps ensure that experiences are accessible. You might consider accessibility as falling under the umbrella of inclusive design. Or to be an outcome of inclusive design. So, practicing inclusive design helps ensure that you're practicing privacy by design. And highly accessible experiences are more likely to ensure everyone's privacy, too.

Reference Thoughtful Personas and Archetypes

Developing personas or archetypes for your experience certainly goes a long way toward ensuring that you are creating a privacy-minded experience, but be sure those personas are thoughtful and that they include meaningful research and feedback from diverse members of your experience's prospective community. Whenever possible, when developing your personas, consider inputs from as many sources as possible. That means conducting your research via a diverse array of methods such as survey results, one-on-one interviews, usability testing, customer feedback, and website analytics.

Consider whether persona spectrums may be helpful, too, since, by design, they are intended to help address inclusivity and accessibility issues. Originally developed by Kat Holmes at Microsoft, they provide a more nuanced understanding of people's needs, including how solutions might benefit different people whose needs are permanent, temporary, or situational.[6] (See Figure 9.1.)

---

6 "Making the Digital World More Inclusive," Microsoft, 26 April 2019, https://news.microsoft.com/en-xm/2019/04/26/making-the-digital-world-more-inclusive

Vision

PERMANENT
Blind from birth

TEMPORARY
Recovering from eye surgery

SITUATIONAL
Dealing with screen glare from the sun

FIGURE 9.1
This example of a persona spectrum from Microsoft depicts how a vision issue may be permanent, temporary, or situational. The same solution might help people in all three modes.

Consider our earlier accessibility example and that some experiences have been created with poor accessibility very intentionally. For example, they may present terribly important copy in a very small font against a background with very poor contrast, so as to present some information, technically, but in a way that reveals that the creator hopes people don't actually read it. In the worst cases, this copy may not be visible at all. (Some describe this problem as trying to find a *polar bear* in a snowstorm.) This problematic pattern creates an accessibility issue, of course, for people with faltering eyesight, but it's also intended to obscure important information from everyone.

In the same way that personas can be extended for accessibility purposes to the general population, so can issues with privacy be extended from causing particular problems for a distinct type of user to, potentially, the general population, as well.

NOTE MISMATCH

For more on inclusive design and persona spectrums, see Kat Holmes's book *Mismatch: How Inclusion Shapes Design*, MIT Press.

> **EXERCISE** IMAGINE A PERSONA SPECTRUM FOR PRIVACY
>
> What might be an example of a persona spectrum you could develop that addresses that persona navigating a particular privacy issue? You might consider a particularly vulnerable group of people and how they would be affected by one of the privacy problems you've reviewed already. The severity of the threat might be a permanent concern for them. Now, think about how an issue for someone in that specific group could also be meaningful for someone in a temporary or situational state.

Employ User Journeys and Stories

To introduce thoughtful privacy features into any experience you'll want to consider user journeys and stories, too. You can use your personas or archetypes as guides to drive them.

A *user journey* is a visual or diagrammatic representation of a user's overall movement within an experience from beginning to end. You could consider different journeys for different personas, since those individuals would have different needs and concerns. Developing a user journey is a fun, creative exercise, but it will also be tremendously valuable in pinpointing moments in an experience that prove to be pain points for them—including privacy issues. Then you can determine how best you can provide a digital solution or feature to address each issue.

Once you've highlighted these issues, you could then translate them into specific user stories that you can address in future design sprints. If you're working within an Agile methodology, those stories might roll up into a privacy-oriented "epic," as well. Each user story should address a specific issue, such as "As a user, I want to make my profile private, so only my friends can see it" or "As a user, I want to know why I should submit my phone number to continue signing up for the experience." If you're working on a smaller project, you could also create a privacy category for your backlog and make it a goal to ensure it doesn't get neglected.

Evaluate Experiences for Harm

If you have ever attended a conference on human rights or internet freedom, you may have interacted with people who had very distinct privacy issues when appearing in public. They may not offer their real names. They may insist on not being photographed. These individuals likely have different backgrounds that prompt them to consider specific harms that could come to them for appearing at such an event. They're living within a very specific state that may differ from yours or mine—despite the fact that they're in the same physical space. It's not unusual for photography to be forbidden at such meetups in certain places and only with explicit consent in other spaces.

Whether they realize it or not, however, *anyone* can be subjected to different levels of harm both online and in real life or "meatspace," even if they've never taken the time to consider it.

To better understand the impact of an experience upon members of a diverse user base, you need to consider the harm aspects it could bring to different people.

If you're working in a space such as human rights or advocacy as a designer, there's a good chance you already consider "harm" as part of your design practice. If you're not working in such a space, you may not think to utilize some relevant techniques and tools, unless you're focusing specifically on cyber security. But thinking about potential harm is instrumental in designing for privacy or, indeed, in creating any sort of safe, thoughtful experiences. The concept of "harm" is significant enough that experts have established risk evaluation practices that you can reference to help analyze whether harm might emerge from the experiences you're designing. Considering harm will prove invaluable for identifying privacy issues and for developing forms of online protection, too, such as blocking and reporting on a social media platform, for example.

Threat Models

Threat models examine the dynamics of any given system from a security perspective to identify weak points that may enable threats to that system—or cause harm. Once threats are identified, the owners of the system consider ways to mitigate them—to develop countermeasures—and then to validate whether those countermeasures have proven successful. It's common for companies to develop threat models to protect themselves against vulnerabilities within their large and complex IT systems. These threats tend to have a more *external* focus. Who might attack us and where?

Though they vary in exactly how they're articulated, you'll typically find that threat models include the following key steps in some variation or another. I've framed these to reflect the experience design perspective:

1. **Define safety and security requirements.** Consider specifically what it is that your experience needs to protect.
2. **Visualize the system.** This may mean creating a diagram depicting the various parts of a system and where data is being handled. You could also examine an existing live experience together to discuss those moments where data is shared and how it's handled.
3. **Identify threats and vulnerabilities.** Now identify and highlight those specific moments where weaknesses or even attacks might occur.
4. **Design and deploy mitigating solutions.** Now you're integrating design solutions to respond to particular issues. For the purposes of privacy protection, these might entail new or more refined settings or the development of new features that ensure privacy.
5. **Validate and iterate upon solutions.** Once those mitigating solutions have been launched, you have the opportunity to test them and to secure feedback from users, as well. (Hopefully, you had the opportunity to do this before launching, too!) Then you can iterate upon the features you've deployed.

The following diagram shows a typical framework/methodology for identifying and mitigating threats within a system (see Figure 9.2).

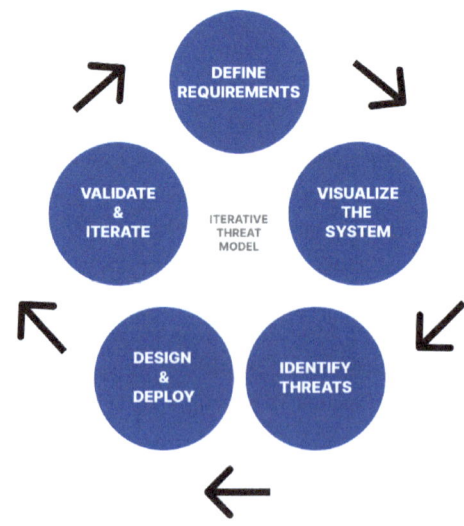

FIGURE 9.2
The five repeatable steps of a threat model.

Similarly, threat models can be used to identify vulnerabilities that may lead to harms that affect people, including at-risk individuals or groups, whether they are human rights advocates, journalists, members of the LGTBQIA+ community, or the average person with their needs for personal privacy and security online. Just as a financial system—with its countless profiles and accounts, transactional features, and so on—can benefit from a thoughtful, detailed threat model, so too can an individual with their own interactions—both online and off—reflecting their particular concerns for privacy and security. That means that, yes, *you* have a threat model, too. Additionally, although you may reference specific at-risk groups when considering an experience for harm, remember that addressing any privacy or safety issues uncovered for *them* ensures that you're addressing them for *everyone* using an experience.

For your purposes, threat modeling proves invaluable for considering possible privacy threats that users might encounter within a system, ideating upon design solutions to mitigate those threats, and validating those solutions after they've been deployed. You can enlist personas or archetypes to help you consider specific threat models, too, such as the needs of a gay person living in a specific country or a journalist working under a repressive regime or a women's rights advocate protesting in a specific region.

Consider, for example, a social media platform that plans to enact a "real name" policy with the belief that insisting that users post under their real names will prompt them to post more responsibly. A thorough consideration of the threat models for various types of users would reveal that insisting on real names for all users would prove disastrous for some, who have a legitimate need for anonymity. To avoid harm, they may just abandon the platform.

Some questions you can ask as you're considering the experience:

- Are we requesting information that could infringe on the privacy of users? (For example, are data points like real name, gender, or location really needed?)
- Will any of this information be visible to other people operating within the system? (It's especially important for social media and collaboration platforms.)

- Will users feel they have access to adequate tools to control how their information is being shared, not just with third parties, but potentially, publicly? (For example, to hide themselves from other users, selectively if need be.)
- Where might users encounter confusion that might cause them to accidentally reveal information they didn't intend to? (For example, perhaps they entered information they didn't realize would be publicly available on their profile.)

Anxiety Games

One quick way to identify potential threats is via an exercise the designer Andrew Lovett-Barron has called *Anxiety Games*. Andrew and his wife Ayla originally developed this game as a way of determining their preparedness for the COVID pandemic.[7] The human-rights centered design firm Superbloom codified this game into a framework useful for considering threats in digital design. Participants are encouraged to consider specific threats that fall into four primary categories: Physical, Digital, Social/Political, and Environmental.[8]

You could conduct an exercise like this in person using Post-it Notes in real life or online using a collaboration tool like FigJam, Miro, or Mural, so your team members can all contribute. (See Figure 9.3.) This is a good moment to consider that there are no wrong answers: You never know what someone may bring up that might lead to a potentially illuminating discussion.

Finally, consider convening a cross-functional team that would conduct this sort of harm analysis *early* in the design process, *not* when work on the product is almost wrapped. That not only enables you to ideate on solutions for any weak spots as you design features, but it also prevents you from last-minute panic moments that may prevent you from launching on time—or, worse, launching a product that may cause harm.

---

[7] Andrew Lovett-Barron "Covid Anxiety: A Foresight Game," 16 May 2021, https://andrewlb.com/blog/covid-anxiety

[8] "Anxiety Games," Superbloom, https://superbloom.design/resources/anxiety-games-template.pdf

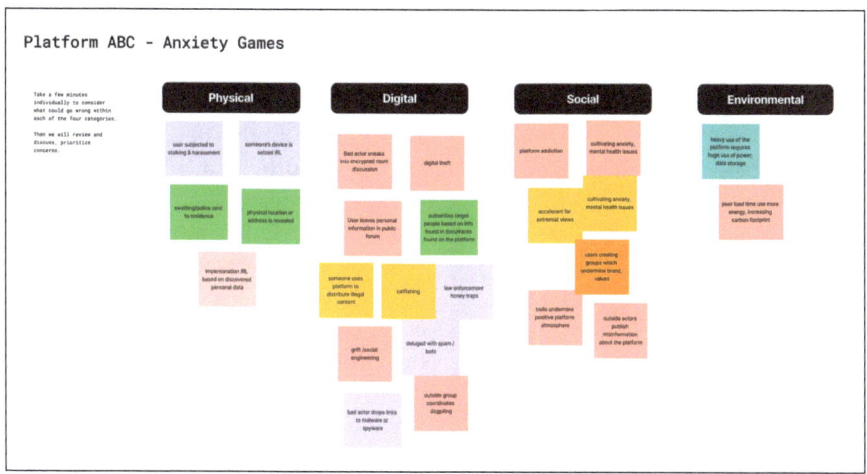

FIGURE 9.3

Example of an Anxiety Games exercise conducted for a platform. Participants have placed potential threats on sticky notes within FigJam under the four provided categories.

WHO'S LISTENING IN?

I was stretching at the gym a few months ago, listening to my workout playlist, when suddenly, something odd happened. I got an alert on my Spotify app informing me that "<User name redacted> is listening on Red Tamales" with options for me to "Join Session" or "Not Now." (See Figure 9.4.) As I recall, there was only one other person working out in the room, and I saw no indication that she had invited me to join her in listening to her Spotify. I assume, in fact, she had no idea that I had gotten the alert. I didn't join her listening session, but her Spotify handle did appear to use her first name along with her last initial or two. Fortunately, that abbreviated version of her legal name helped somewhat to obscure her full identity because this was still a significant privacy violation—similar to that of Strava's. (I did, however, find screenshots online posted by other users that appeared to reveal the entire name of the person unwittingly sharing their listening session.)

continues

WHO'S LISTENING IN? (continued)

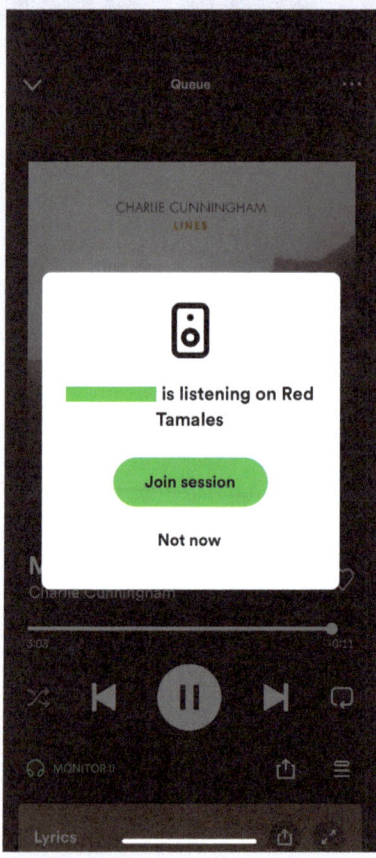

FIGURE 9.4
This Spotify alert includes a prompt to join the user's listening session. Her username has been edited out.

Now, Spotify may well have emailed everyone about this new feature called *Jam* or even provided an alert within the app to explain it at some point that I didn't notice. Either way, however, as a heavy user of the app, I hadn't heard of the feature, so I suspect many, perhaps most, others hadn't either. And yet, this Premium subscribers' feature was launched "on" by default, so it allowed anyone with a Spotify app nearby to join this user's session. A little investigation revealed that this prompt was likely triggered because we were on the same Wi-Fi network at our gym. I experienced the same issue a few times elsewhere. In fact, I discovered, people had complained directly to Spotify that if they were on a hotel or apartment building Wi-Fi network, their listening session could be shared with dozens of people. (See Figure 9.5.)

In addition to potentially sharing someone's identity, however, there are additional privacy concerns with this sharing model: Spotify features not only a vast array of music of all styles and genres, but myriad books and podcasts, too. A user could have any number of reasons for wanting to keep what they're listening to private. Imagine a woman is pregnant and hasn't told anyone, and she's listening to a podcast about pregnancy. Imagine a young gay or transgender person is listening to a podcast about LGTBQIA+ experiences as they are grappling with their own sexuality or gender identity. Imagine someone has been recently diagnosed with an incurable disease and is researching it. Or imagine someone is simply listening to some fairly explicit material, and they never imagined anyone else could be tuning in, too.

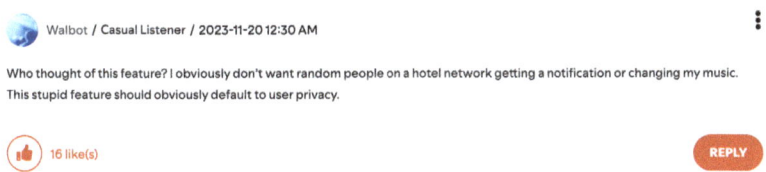

FIGURE 9.5
A post to the Spotify feedback community notes these notifications as a privacy issue and emphasizes the need for privacy as a default.

Additionally, responding to questions and criticism in a feedback forum, a Spotify employee confirmed that these notifications could *not* be turned off at the time. Many users complained about the privacy issues with the feature, including one subscriber Lejos, who said, "Automatically searching for users on the same Wi-Fi seems crazy intrusive to me what if we're on a public Wi-Fi at a mall? How many prompts would you be receiving to join?" Over two years later, people were still posting in the forum to confirm that the issue had not been addressed. By the time of this writing, 432 people had voted on one of three threads I found to say they would like to see the issue addressed. Seemingly then, Spotify does not have a design culture that focuses specifically on privacy by design; otherwise, they would have made fixing this issue a high priority.

Take a moment to consider several issues with Spotify's Jam feature as it was deployed:

- This sharing feature was launched "on" by default for people accessing the same Wi-Fi—and still has not been updated to allow users to turn it off.
- It invites people to the listening session without the subscriber's explicit consent.
- People could listen in without the host knowing.
- It could expose someone's identity.
- It could expose their personal listening material.

If pressed to rate this experience from a privacy perspective, I'd have to give it an "F." Assuming the feature does provide some value with specific use cases, a more concerted effort would have to be made to design the overall user experience with everyone's privacy in mind—especially as users move in and out of different environments that may be public or private.

Drive Change as a Design Leader

If you are a design leader or manager, you're well positioned to help nurture a team culture that's more likely to keep privacy in mind throughout the design process. You have the distinct opportunity to cultivate a team culture that respects and listens to all contributors and to discourage a design culture that caters only to one or two more vocal "rockstars." You can look for those opportunities to incorporate more participatory activities, to run that collaborative sketching exercise, and, of course, to consider privacy issues, specifically.

Top-Down Buddies

A network of support helps, too, especially when that support is both modeled and provided from the top down. In her book on policy and design, *Deliberate Intervention*, Alexandra Schmidt suggests that designers will feel more empowered to speak up against deceptive patterns, for example, if they benefit from a "top-down buddy."[9] That individual would be a manager or executive, who they're confident will support them in making such a stand. Such mentor roles are instrumental for cultivating a culture of privacy by design. The opposite dynamic could cultivate a chilling effect, where employees are instead afraid to speak up if their manager or another superior, let's say, actively advocates for deceptive patterns because they increase profits.

I'm happy to say that I benefitted tremendously in my career as a UXer from a manager or two, in particular, who acted as effective top-down buddies. They not only helped me feel empowered to advocate for better privacy in design, but also to pursue my interest by speaking and writing on the topic and to further educate myself by attending conferences and networking with people and organizations I could both learn from and assist. (Consequently, you'll find a couple of these individuals noted in the acknowledgments for this book.) Additionally, they instilled a real desire in me to pass on that sort of support myself—to ensure that I, too, became a top-down buddy. Tangentially, too, I've come to believe, this is how resilient, meaningful support networks are built and maintained: One relationship at a time.

---

9 Alexandra Schmidt, *Deliberate Intervention: Using Policy and Design to Blunt the Harms of New Technology* (Two Waves Books, 2023).

Never Stop Learning

Finally, while there's a lot of discussion within the field of user experience design around what tools to use (Figma, Mural, etc.) or the right methodologies to follow (Agile, Lean UX, Design Thinking, etc.), you shouldn't ignore the role of developing knowledge in areas that may seem tangential to your "core" qualifications but that do prove helpful in learning to work in ways that are more conducive to practicing privacy by design.

You might study up on skills like interviewing or conducting surveys that help you get to know your audience better. But you might also tackle topics you didn't previously think even related to your work. Consider broader topics like diversity and inclusion, certainly, but also topics that focus on the very specific needs of different identity groups. Remember that considering the needs of these specific groups invariably leads to creating a safer experience for everyone, too.

If you're creating products that will be used outside of the country or region you're living in, it's vitally important for you to be aware of how people's concerns about specific issues vary globally, too. For example, as you saw with Grindr in Egypt, a dating app with more than 14.5 million monthly active users can bring gay people together in one part of the world with relative safety, but can be used to ensnare and endanger them elsewhere.

Here's a nonexhaustive listing of topics you might investigate that could further your understanding of people's needs when designing for privacy:

- Cennydd Bowles on "Future Ethics"
- Speculative design, including probability cones and futures wheels
- Ethical use of artificial intelligence
- Cross-cultural product design
- Value sensitive design
- Prosocial design
- Risk models for journalists and human rights workers
- Security concerns for the global LGTBQIA+ community

If a list like this seems intimidating, don't fret: You can't educate yourself on all these topics at once, but it does become particularly important for you to study up on particular issues if the projects you're working on touch on them. If you're a design leader, you could find ways to introduce these topics to your team, perhaps within a weekly meeting via a short presentation or by inviting in a relevant guest speaker.

> **EXERCISE** A SAFER SPOTIFY
>
> Take a moment to recall the preceding example of Spotify's Jam session and its potential privacy issues. What advice would you give to a client, who planned to debut a similar feature?
>
> Some questions for you to consider:
>
> - What different audiences or personas for the Spotify app may be operating under different threat models?
> - What specific harms might concern different personas operating within the experience?
> - Who might you include in brainstorming or usability sessions to mitigate harm-causing vulnerabilities being built into this experience?
> - How might you roll out a feature like this in a way that is respectful to everyone's privacy, but still makes it visible and available for those who would like to use it?

The Takeaway

You've just skated through a few different topics, any one of which you could explore in much more detail. Consider this chapter a brief overview of some important tools and practices that can contribute to developing a culture conducive to designing privacy-centric experiences. Whether you're studying up on accessibility, inclusive design, persona spectrums, or harm evaluation, you'll be learning methods and best practices that will help you become a more thoughtful and effective user experience designer, as well.

And if you're a design leader, you should feel empowered now to be more assertive in cultivating teams that practice privacy by design.

CHAPTER 10

Exercises: Navigating Two Privacy Scenarios

Scenario 1: Friends, a Networking Site   203
Scenario 2: BookLuvr, a Book-Sharing App   204
The Takeaway   206

Having gotten to this point in reading this tome, you've probably grasped the fact that, like seeing a mouse in your house, if you find a privacy problem within an experience, there's probably a few more of the critters running around in the walls. You may well just encounter a simple issue like an empty expectant field requesting personal information that doesn't need to appear on a form or could just be made optional. There's a good chance, however, that if you're working within an organization that doesn't practice privacy by design, you're likely to discover privacy issues as systemic, possibly even intentionally architected throughout the experience. Either way, these organizations will benefit from your help.

As a little mental and creative exercise, take a look now at the following two fictional scenarios that depict projects characterized by problematic privacy issues that could prove harmful both to users *and* the companies putting them out there. Think about how you might address these issues as a designer. Consider some of the following strategies you've reviewed and determine how you might engage them in each scenario:

- Initiate conservations with stakeholders and team members to explain the privacy issues.
- Conduct research to determine the privacy needs of a diverse set of users.
- Consider carefully what personal data is really needed and how it's being used.
- Use clear language to describe how data is being used and why.
- Avoid deceptive patterns and consider best practices for creating better design patterns.
- Prioritize the design and deployment of helpful privacy tools and features for users within the experience.

If you would like to make the most of these exercises, take some time to sit somewhere—at home, at your favorite coffee shop—with your laptop or some scratch paper. Write down your ideas for handling the issues that arise from these scenarios and sketch out some potential design solutions. Consider how you might pull together a short but effective presentation to make the case for your solutions.

Scenario 1: Friends, a Networking Site

High-level issues: *Use of deceptive patterns, vague, impenetrable privacy language as a strategy for growing a user base, and misuse of sensitive user data.*

As a UXer with several years of experience, you've just been hired to provide user experience oversight and design for a new professional networking site called *Friends* that already has a lot of work in motion, although the platform has yet to make its public debut. Once launched, Friends hopes to build an engaged audience quickly and exponentially via a strategy of "growth hacking" and forced invites. They plan to ask users to provide access to their contact lists, so they can find their friends, who have already signed up for the site, but, on an already crowded key screen, they also include a brief sentence in smaller text that says, "If we don't find your friend on here, we'll send them an email encouraging them to sign up!"

Additionally, nothing in the designs underway explains prominently how the company plans to share users' data with dozens of third parties as a means of generating income. The content that does tackle this subject is squirreled away deep within the platform's still evolving Privacy Policy, and it's presented there within big blocks of unformatted text and written with language that's very vague, generalized, and lawyerly.

You note, too, that setting up a profile on Friends requires users to provide the fledgling company with a lot of personal information, especially about their personal preferences, ostensibly to help users discover new friends. When you ask the founder whether they would share user data with authorities, if asked, the owner shrugs and says, "Of course." You're concerned that the authorities in some places may misuse the very detailed and private data collected here for profiling.

What advice and direction could you give this company? What changes would you advise specifically be made to the platform before it's released into the wild?

Need a nudge or two to jump-start your thinking? Here are some ideas:

- Consider developing an inventory of personal data points requested across the experience to determine whether they are all needed and what their potential privacy impacts may be.
- Propose restructuring the privacy section of the site to highlight specific privacy issues users should be aware of.

- Propose onboarding to explain to users how their data will be used and which third parties it will be shared with.
- Examine any design patterns carefully to ensure that they don't trick users into sharing their contacts or otherwise pressure them into sharing information in the pursuit of growth hacking.
- Ensure that users have a way to import their contacts if they want to but can easily opt out of spamming all of their contacts to ask them to join.
- Brainstorm more creative and transparent ways to grow your user base.
- Conduct usability testing around the sign-up process to learn what new users feel comfortable providing and what will create hurdles for them, possibly even prompt them to abandon sign-up.
- Consider whom to approach to discuss your concerns about the potential harm that can arise when sharing users' data with authorities, unless there's a particularly compelling reason to do so.

Scenario 2: BookLuvr, a Book-Sharing App

High-level issues: *Accidental stalkerware, lack of diversity in design culture, no insight into potential for harm, and public by default data sharing.*

You're a junior designer who has been working for a fun book-sharing app created by a small startup company call *BookLuvr*. You love the culture, and as a book lover yourself, you believe in the product. Your new job hardly seems like work!

To kick off the quarter, your team just held a brainstorming meeting to discuss new ideas to promote engagement on the app. The most popular idea to come out of the meeting was a proposed new feature that would alert users when someone is nearby, who has similar tastes and reading patterns as you. The experience would share the stranger's favorite books with you as a way of introducing you both, as well as their profile photo and handle. This feature would be on by default upon rollout because the founder believes it will be fun and "a great way to meet new people, maybe even like a dating app!" The app will need to encourage users to allow location tracking constantly within the app, too.

You leave the meeting feeling a little uneasy. You briefly asked whether this feature might present some privacy concerns but left it at that. Back at your seat, you decide it sounds like a terrible idea. But what can you do now? You're new. Maybe you feel a little reluctant to make waves. Have you lost your opportunity to have any impact on the feature and how it's implemented from a privacy perspective? What would you do next?

Need a nudge or two to jump-start your thinking? Here are some ideas:

- Propose a survey of users to determine how valuable they might find such a feature and what their privacy concerns might be.
- Conduct an internal threat-modeling exercise to evaluate potential harm across personas or archetypes likely to be using the app. You might try a quick session of "Anxiety Games" online or in person as described in Chapter 9, "Cultivating a Culture for Privacy by Design."
- Propose a detailed educational rollout, including onboarding that explains the benefit of using the feature to users, but asks their permission to turn it on first.
- Ensure that both existing and new users are alerted to the new feature by emailing users about it and providing just-in-time alerts for existing users, too.
- Propose usability testing upon the designs once they've been completed, but before they've been developed and pushed live to highlight any issues that may arise from a usability and accessibility perspective, but also from a privacy perspective.
- Take your thoughts to your manager to reiterate your concerns with and suggestions for the feature and to enlist his or her help in either pushing back on the feature or steering it in a different direction.
- Consider putting a brief presentation together and ask for time with the design team to discuss the issues as you see them, propose alternatives, and discuss potential next steps.

The Takeaway

How did you do? Hopefully, you're excited to find at this point that you have quite a few best practices, tools, and helpful new design patterns at your disposal to make a compelling case for designing for privacy in these scenarios. It would be worth your time to take a few minutes to sketch out a framework or user flow depicting your solutions for each scenario, just to get you envisioning not just a broad approach to these issues, but also to explore the details of how you would handle them.

Are you an entry level designer looking to show your chops? You could consider a real-life company with known privacy issues and work up a case study that details how you'd solve those issues, explaining how you'd handle the scenario, what best practices and tools you might engage, and showing mockups for your improved designs. You could even include this output in your portfolio.

If you're a design leader, you might consider conducting these exercises with your team to lead a discussion around the value of privacy by design. You could use an online collaboration tool to generate feedback and ideas and host the exercise during a periodic "Design Jam" session. You could make it a cross-disciplines activity with your colleagues, too. You might discover that some of the output proves valuable in surfacing helpful changes or features for your own products.

CHAPTER 11

The Evolving Impact of Privacy Policy

The Impact of Europe's GDPR	208
The California Consumer Privacy Act	211
The Rest of the United States	214
Evolving Global Efforts	215
Enforcing Penalties	217
Corporate Self-Regulation and Policy	218
Massive Companies Enforcing Privacy Unilaterally	219
The Takeaway	224

For several years now, emerging international regulations have been forcing companies to pay more attention to privacy issues, a trend that seems only to be growing, as people learn more about the value of maintaining control over their own personal data.

I've spoken with colleagues in the past who were concerned that we can't make any traction on privacy issues without government intervention via policies and regulations. I've also spoken with some, perhaps of a more libertarian bent, who question whether government intervention can help with privacy issues, at all, perhaps pointing to ineffective legislation that has also had unintended, sometimes damaging consequences.

We can look, however, to a long history of policy intervention that has created a demonstrably safer environment for people, whether it's by removing asbestos from building materials or insisting that seatbelts and airbags be added to automobiles, removing mercury from car switches, banning lead from cosmetics, paint, and gasoline, and so on.

You've likely heard the observation usually attributed to Winston Churchill: "Democracy is the worst form of government except for all those other forms that have been tried from time to time." Similarly, the policies we implement as a society may not be perfect, but, fortunately, just like any other form of design, good policy design is iterative. Policies can be updated and improved along the way, introducing nuance where it's needed or removing those elements that prove to have unintended damaging consequences.

Indeed, that's exactly what we see happening with the privacy-oriented policies that governments are debuting and then iterating upon around the world.

The Impact of Europe's GDPR

Remember a few years ago, when you suddenly received a deluge of emails from companies globally informing you that they had updated their privacy policies? That rush on your inbox was a direct result of the European Union's General Data Protection Regulation (GDPR) going into effect.

A more general right to privacy was established as a human right at the 1950 European Convention on Human Rights, and various other laws and regulations around privacy have been established in Europe since then, laying the groundwork for the GDPR. In 1995, for example, the EU passed the Data Protection Directive—a year after the first banner ad debuted online. Then in 2011, with online life having evolved tremendously, Europe's data protection authority decided that a more detailed framework for considering privacy needed to be developed. That became the GDPR.

Well known to anybody doing business online now, the GDPR was finalized in 2016 and officially came into effect on May 25, 2018. This landmark set of privacy regulations continues to be the framework making the greatest global impact, despite having been formulated for the EU. Why?

The GDPR regulates how businesses can gather and transfer or process consumer's personal data online when they are working within the European Union. However, those companies do not have to be based in the EU. These regulations apply, regardless of where a company exists, if they are processing the data of consumers living in the EU. The GDPR also regulates what happens to that data when it's transferred outside of the EU.

So, yes, if you're a U.S. company with a European audience and you're offering goods or services—whether money is exchanged or not—or if you're tracking the behavior of people within EU member countries (noting IP addresses, dropping cookies), you have to adhere to the GDPR. There's an exception for companies with fewer than 250 employees in that they don't have to keep records of their data activities; however, they can still be held responsible for privacy violations if, for example, they're processing data that presents a risk to the rights of individuals or that includes information about people's health, ethnic origins, or criminal convictions.

Note some of the more significant requirements of the GDPR:

- Companies must ask consumers to opt in to sharing their data.
- They must communicate to consumers in the moment when they are collecting their personal data.
- They must be transparent about what they are doing with that data.
- They must provide individual consumers with the ability to both download their personal data and to delete it, a practice often referred to as "a right to erasure" or "a right to be forgotten."

The ubiquity of cookie banners is a highly visible outcome of the GDPR, too. An unintended, somewhat ironic consequence of that is that you're often presented with cookie banners that incorporate deceptive language or patterns, positioned to hasten your permission to drop as many cookies onto your devices as possible. Still, you can assume that future amendments to the GDPR and other emerging regulations will require more refined and transparent versions of this feature—if cookie banners remain a viable solution to this particular privacy issue at all.

Many Silicon Valley companies were not happy with the advent of the GDPR in 2016. Many companies bristled at the regulations that would presumably block them from processing data in ways that had enabled them to generate significant profits. Still, those companies had the scale and largesse to devise new ways of creating profits or to circumvent the regulations. Google, for example, simply foisted off responsibility for managing consent on any partners, publishers, or advertisers that subscribed to their technology or ad targeting solutions, such as AdSense or AdMob. And Facebook hired battalions of lawyers, lobbyists, and managers to respond to the GDPR. By 2022, a study by the Centre for Economic Policy Research showed that GDRP did prompt an average loss in profits for companies of about eight percent and a two percent reduction in sales.[1] These losses were likely incurred due to compliance costs: Companies had to spend money updating their technology, for example, and the costs for obtaining consent for collecting data rose. However, these setbacks primarily hit small and medium-size companies, and large companies like Apple, Facebook, and Google did not appear to have suffered any losses at all. In fact, all three of those companies expanded their customer bases in that time, likely even benefiting from those smaller companies' inability to compete in this new environment.

Nonetheless, from a consumer perspective, implementation of the GDPR has proven modestly successful. By May 1, 2025, the accumulated fines levied via the GDPR alone were well over $6.4 billion.[2] Over 2,300 violations contributed to that total, including $1.3 billion coming from a single settlement against Meta.

---

1   Giorgio Presidente and Carl Benedikt Frey, "The GDPR Effect: How Data Privacy Regulation Shaped Firm Performance Globally, Centre for Economic Policy Research, 10 March 2022, https://cepr.org/voxeu/columns/gdpr-effect-how-data-privacy-regulation-shaped-firm-performance-globally

2   GDPR Enforcement Tracker, accessed 1 May 2025, www.enforcementtracker.com/?insights

Despite its imperfections and the ability for companies to wriggle their way through its restrictions, however, the greatest success of the GDPR may go unseen: It has demonstrably improved privacy online for millions of people.

The California Consumer Privacy Act

In the United States, California leads the way with online privacy legislation. In 2018, California passed their own version of the GDPR: The California Consumer Privacy Act or CCPA.

The CCPA gives Californians more control over how their personal data is used, and, at a glance, its requirements look very similar to those in the GDPR. However, one important distinction between the two laws is that the CCPA allows businesses to collect consumers' information *by default*—although they still have to offer the ability to opt out. (Perhaps not a surprising distinction between regulations originating within the United States versus Europe, respectively?) Also, businesses that handle data for fewer than 50,000 users are not subjected to the regulations.

The CCPA has resulted in a $1.2 million fine against Sephora, the cosmetics retailer because they failed to disclose to consumers that their data was being sold to third parties. Sephora also failed to address user requests to opt out of data sharing.

In 2020, California voters agreed to update and strengthen the CCPA with the California Privacy Rights Act, sometimes called *CCPA 2.0*, which went into effect in 2023. Among other things, it further enabled Californians to update inaccurate personal data and limit businesses' use of their "sensitive personal information."

Arguably, the CCPA has proven successful in that it has prompted other states to follow suit with their own privacy laws, while also raising awareness of privacy issues in the U.S. more generally. Critics, however, also note that enforcement has been slow and the costs for compliance have been high for small and medium-size companies.[3]

---

3 Sari Richmond, "Effectiveness and Implications of the California Consumer Privacy Act," *Northwestern Undergraduate Law Journal*, 25 August 2024, www.thenulj.com/nuljforum/effectiveness-and-implications-of-the-california-consumer-privacy-act

Banning Deceptive Patterns

In March 2021, California announced a law banning "dark" or deceptive patterns, the first of its kind to succeed in passing in the United States.

Earlier, in 2019, two senators—Virginia's Mark Warner, a Democrat, and Deb Fischer, a Nebraska Republican—introduced a bill that would have banned deceptive patterns from any online platform with over 100 million users. The Deceptive Experiences To Online Users Reduction (DETOUR) Act stalled and, ultimately, was never voted on. Nonetheless, Senator Warner and colleagues reintroduced the bill in 2021 and 2023, so there's hope that some incarnation of the bill will pass, eventually.

Smartly, California's updated regulations include specific examples of deceptive patterns and explain how certain privacy-related features must be handled.

CALIFORNIA FORBIDS DECEPTIVE PATTERNS

Some examples of the deceptive patterns California has banned, taken directly from the state's updated privacy act regulation text:[4]

- A business shall not use confusing language, such as double-negatives (e.g., "Don't Not Sell My Personal Information"), when providing consumers the choice to opt-out.
- Except as permitted by these regulations, a business shall not require consumers to click through or listen to reasons why they should not submit a request to opt-out before confirming their request.
- The business's process for submitting a request to opt-out shall not require the consumer to provide personal information that is not necessary to implement the request.
- Upon clicking the "Do Not Sell My Personal Information" link, the business shall not require the consumer to search or scroll through the text of a privacy policy or similar document or webpage to locate the mechanism for submitting a request to opt-out.

---

4 Final Regulation Text Title 11, Law Division 1, Attorney General, Chapter 20, California Consumer Privacy Act Regulations, Office of the Attorney General, 2024, https://oag.ca.gov/system/files/attachments/press-docs/CCPA%20March%2015%20Regs.pdf

Once California determines that a banned pattern is being used, the company involved has 30 days to remove it or face fines.

This law serves as a helpful precedent for banning specific patterns, creating a list or taxonomy of deceptive patterns that can scale as more patterns are identified and determined to be harmful.

California's Privacy Options Icon

Along with this update around deceptive patterns, California also debuted a new "Privacy Options" icon for companies to show users where to look if they want to opt out of their data being collected and sold by businesses. Carnegie Mellon's CyLab Security and Privacy Institute and the University of Michigan's School of Information collaborated to design the icon (see Figure 11.1). It's intended to be used at a size that matches any other icons used on a company's site. The icon was soon added to many sites belonging to companies like Ford Motor Company, Procter & Gamble, Spotify, Verizon, and Walmart. In context, its appearance may seem modest, but it does draw attention to important content and features that often get overlooked in website footers—sometimes by design.

FIGURE 11.1
You may find California's "Privacy Options" icon on websites now. It highlights features that enable users to opt out of having their data sold.

The Delete Act

In October 2023, California's Governor Gavin Newsom also signed the Delete Act into law. This bill enabled California residents to request that their personal data be deleted from any data broker's databases. These companies had to register with the state government and provide a single simple form allowing individuals to request deletion from every single company at once. This was the first bill in the United States to enable such universal data deletion, and, unsurprisingly, not everyone was happy about it.

One advertising industry group lamented the Delete Act, suggesting that companies that rely on third-party data would suffer financially. They also argued that the Delete Act would empower specific tech companies like Apple, Google, and Meta that benefit from controlling users' data within their own walled gardens. Human rights and privacy groups, however, applauded the Act, since, for example, it had been previously reported that organizations such as Immigration and Customs Enforcement (ICE) had used such data brokers to secure data about individuals to circumvent some place's sanctuary laws in efforts to highlight people for deportation.

Oregon, Texas, and Vermont have also passed laws creating databroker registries. Vermont's law launched in 2019, Texas's in 2023, and Oregon started mandatory registration on the 1st of January 2024.

The Rest of the United States

If California provides the most detailed example of a state's privacy laws in the United States, as of April 2025, 20 U.S. states had debuted some sort of detailed online consumer privacy laws. Encapsulated in the Florida Digital Bill of Rights (FDBR), Florida's new data privacy laws are among the more stringent of these, requiring Florida-based companies to get an explicit thumbs-up from users before selling their data to third parties. The bill also forbids the use of deceptive patterns and gives particular focus to protecting children's personal data from collection, sale, or storage.

The State Data Privacy Act offers an interesting model, too.[5] Created by the Electronic Privacy Information Center (EPIC) and Consumer Reports, this bill is based upon the Connecticut Data Privacy Act (CTDPA), since it was considered an established law recognized by lawmakers, despite its perceived flaws. Additionally, the text from Connecticut's bill was also updated to close some loopholes and to provide greater emphasis upon data minimization, civil rights, and private right of action.

The Act's stated goals are to:
- Limit ubiquitous online tracking.
- Encourage more privacy-protective methods of online advertising.

---

5 Electronic Privacy Information Center and Consumer Reports, The State Data Privacy Act, EPIC, September 2024, https://epic.org/wp-content/uploads/2024/09/State-Data-Privacy-Act-EPIC-CR.pdf

- Protect the most sensitive data, including data about kids and teens.
- Use language from existing state laws.
- Allow for meaningful enforcement of the law to ensure compliance.

The Act aims to strike a balance between the needs of consumers and the desire for smaller businesses to protect themselves. It provides a ready-made solution for states wanting to pass comprehensive privacy legislation that they can still tailor to their specific needs. It's also supported by organizations such as The Center for Democracy & Technology (CDT).

Some people will complain that this patchwork of varying state privacy laws makes corporate compliance increasingly burdensome. Nonetheless, these changes generally do make for meaningful advances for people's privacy—and they make a compelling case for practicing privacy by design. Eventually, if this welter of state privacy laws becomes too unnavigable, that may finally prompt the formulation of comprehensive federal privacy laws—something the United States currently lacks.

Evolving Global Efforts

Many other countries have introduced their own, sometimes strict, online privacy laws, too. Take a look at the laws developing in three countries representing completely different parts of the world—Australia, Brazil, and South Korea—while keeping in mind that such policies continue to appear and evolve in the overwhelming majority of countries. In fact, as of early 2024, 137 countries had national data privacy laws in place, accounting for almost 80 percent of the planet's population.[6]

Often, the laws of these different countries overlap, or at least they may not conflict with one another. However, if the laws of a particular country prove too pressing for some companies, they may simply choose not to operate there.

---

6 Aly Apacible-Bernardo and Luke Fischer, "Identifying Global Privacy Laws, Relevant DPAs," IAPP, 19 March 2024, https://iapp.org/news/a/identifying-global-privacy-laws-relevant-dpas

Australia

Australia instituted its first privacy legislation with the Privacy Act in 1988, which intended to regulate how the Australian government, as well as larger companies and organizations, could handle an individual's personal information. Since then, the law has been amended with attempts to regulate the online collection, use, storage and disclosure of personal data by both the public and private sectors. In 2024, the Australian government introduced a bill that it hoped would bring these laws further into the 21st century. It also planned to introduce a children's privacy code, as well as tiered penalties for failing to comply with the law. A prominent case unfolding there as of this writing in 2025 involves Medibank, which is being sued under the Privacy Act for the way it handled data for some 9.7 million Australians. Given the Act's policy had allowed for fines of up to A$2.22 million for each individual affected, the maximum, if unlikely, fine that Medibank could be pressed to pay is a staggering A$21 trillion.

Brazil

Similarly, Brazil instituted their Lei Geral de Proteção de Dados (or LGPD, General Data Protection Law in English). This data protection law regulates the collection, use, processing, and storage of personal data. It applies to any individuals or companies operating in Brazil or any companies outside of the country that process data belonging to Brazilians. This law also established the National Data Protection Authority (ANPD), which has already ordered Meta to stop using Brazilian data to train its artificial intelligence. In October 2024, when Elon Musk announced that X's privacy policy would be updated to confirm that the platform's AI feature Grok would be trained on user-provided content, too, the ANPD said they would seek to determine whether X's policy update met Brazil's standards. In late 2024, a Brazilian consumer rights group, Collective Defense Institute, also announced they were suing Meta, TikTok, and Kwai (another Chinese video app) for failing to warn minors about the dangers of social media use. Their two lawsuits hope to secure a combined 3 billion reais ($525.27 million) from those companies.

South Korea

Known now for having some of the toughest privacy laws on the planet, South Korea passed their Personal Information Protection Act (PIPA) in 2011 and updated it in 2023. Individuals have to provide consent before their data can be processed, and penalties for breaching the law can mean fines and even imprisonment. In 2024, PIPA published guidelines for companies wanting to do business in South Korea to ensure that they complied with the law. Later in 2024, South Korea also fined the cryptocurrency company Worldcoin Foundation 1.1 billion Korean won ($829,000) for failing to adhere to PIPA. Following consumer complaints, a privacy watchdog group concluded that Worldcoin and their developer partner Tools for Humanity had illegally collected biometric data, such as iris scans, from some 30,000 South Koreans.

Enforcing Penalties

What recourse would a governing body have, however, if a company simply refused to pay their fines? For example, let's suppose a U.S. company simply refused to pay fines levied because of breaches to the GDPR? In this case, the GDPR would have several ways they could penalize the company and enforce the fines.

They can do the following:

- Increase the size of the fine, repeatedly, if need be, until it was paid.
- Temporarily or permanently ban the company from processing data (even to the point of banning the app or platform within a specific country).
- Publicly disclose the company's noncompliance to enact reputational damage against the company.
- Conceivably, pursue criminal charges against companies, organizations, or even individuals, especially if the violations were considered particularly egregious.

For fear of violating the GDPR and not having the resources to pay the fines or endure other consequences, some companies have decided to stop doing business in the European Union altogether. Similarly, some gaming platforms have simply blocked EU residents from accessing their platforms rather than rolling out enhanced privacy features or altering their way of doing business. Given the ongoing evolution of privacy regulations globally, however, such companies may find themselves playing in an increasingly smaller space.

Where does all the money from those fines go? In the EU, fines levied via GDPR violations typically go to the relevant national data protection authority (DPA) and then become national funds of the member countries where the fines were issued. In the United States, fines collected by the FTC go into the general fund of the U.S. Treasury. Fines resulting from the California Consumer Privacy Act must be delivered into the special Consumer Privacy Fund to help offset any court costs. You'll find similar examples of how funds are handled in other states and countries, too.

Doubtless, more regulations will be in place globally by the time you read this book. If it means progress toward more meaningful and comprehensive privacy laws, I wouldn't be too upset if you found that this section of my book proved to be somewhat out-of-date when it hit your hands.

Corporate Self-Regulation and Policy

Companies are capable of developing their own policies and standards for privacy, too, by governing themselves instead of awaiting government demands. Businesses can distinguish themselves by self-regulating, given consumers' increasing concerns about the privacy and security of their activity online.

If a particular company is large enough, these restrictions can have a tremendous impact upon other companies' ability to access, process, or distribute your personal data. The largest tech companies, such as Apple, Amazon, Facebook, Google, and Microsoft, all self-regulate to some degree. (That, of course, does not mean they don't feature their own privacy issues.)

For example, Mozilla, most famous for their Firefox browser, promises never to sell your personal data to third-party advertisers. And smaller companies like Lemonade and Lyft also pledge never to sell your data to third parties. The California internet provider Sonic promises not only to never sell your data, but also not to volunteer their users' data to the government or law enforcement for the sake of surveillance.

Additionally, companies can participate in self-regulating organizations established within their industries. For example, the Digital Advertising Alliance (DAA) was created to enforce self-regulating privacy principles across the digital advertising industry.

Companies may also promote the degree to which they regulate themselves. Mozilla, for example, has long made privacy a central theme for both their brand and their software.

Massive Companies Enforcing Privacy Unilaterally

Although they're not creating law, big companies can have a profound impact upon consumers' privacy by mandating how both people and organizations can operate within their increasingly sprawling global platforms. When those companies are as big as Apple and Google, such changes can improve data privacy for hundreds of millions of people.

Apple's App Tracking Transparency

In a sign of the times, Apple debuted an iPhone feature in 2021 called *App Tracking Transparency* (ATT), a privacy forward, anti-tracking shield that prevents apps from stalking you across the internet. (See Figure 11.2.) At least, now, they have to ask for your consent first. Upon release, this feature proved to be immensely popular. By the third quarter of 2021, about 80 percent of iOS users used the new feature to opt out of allowing app tracking for some of the larger social media platforms. That meant the amount of data companies like Facebook, Twitter, and YouTube were collecting via iOS apps to personalize advertising had effectively cratered.

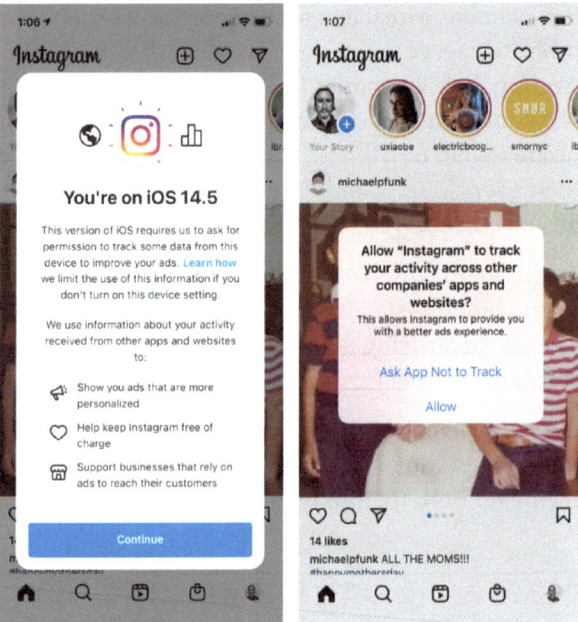

FIGURE 11.2
At left, Instagram's onboarding screen attempts to sell you on app tracking, while, at right, iOS's native messaging more matter-of-factly asks whether you'd like Instagram to follow your travels all across the internet.

By March 2022, however, the global opt-in rate for tracking by iOS users reached 46 percent. By 2024, the global figure appeared to reach about 50 percent of users opting in.

There's some indication, too, that different audiences are more likely to allow tracking. Gamers, for example, tend to opt into tracking at higher rates. So do consumers using finance and utilities apps, presumably because people believe they can better trust companies of their scale.

Similarly, Apple's Location Services feature in iOS also requires companies to request your permission before tracking your movements and geographical location (see Figure 11.3). I found this feature helpful, recently, when iOS prompted me to ask whether I wanted to allow U-Haul to use my location—even when I'm not using the app. Three months after our moving day. I selected "Keep only while using." And I deleted the app. (Although I don't always follow this advice myself, this is a good reason to delete any apps you're not using!)

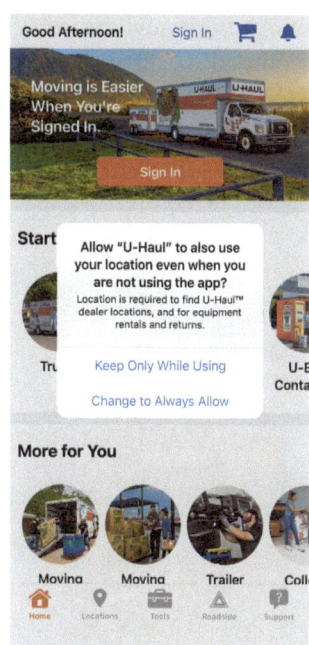

FIGURE 11.3
U-Haul hopes to track my whereabouts every day, not just on moving day.

A wrinkle or two have arisen for Apple, however: In February 2025, a watchdog agency, the Bundeskartellamt (Federal Cartel Office or FCO) believes Apple's ATT framework breaches Germany's antitrust laws, since, they argue, it preferences Apple's products over others'.[7] Further, they argue, the actual design of Apple's related privacy features preferences Apple over others'. In a press release, the FCO concluded, "The current design, in particular the wording, of the dialogue for Apple's own apps makes it more likely that users will consent than that of the ATTF dialogue for third-party apps."

---

7   Natasha Lomas, "Apple's App Tracking Privacy Framework Could Fall Foul of German Antitrust Rules," TechCrunch, 13 February 2025, https://techcrunch.com/2025/02/13/apples-app-tracking-privacy-framework-could-fall-foul-of-german-antitrust-rules/

Then, in March 2025, the French Competition Authority definitively ruled that Apple's ATT framework was anticompetitive during the specific timeframe of April 2021 through July 2023, creating an unnecessarily complex experience for users and "causing definite economic harm to application publishers and advertising service providers."[8] The body fined Apple €150,000,000. (See Figure 11.4.)

FIGURE 11.4

An illustration from the French Competition Authority explaining the fine it levied against Apple for their App Tracking Transparency framework.

---

8 "Targeted Advertising: The Autorité de la Concurrence Imposes a Fine of €150,000,000 on Apple for the Implementation of the App Tracking Transparency ("ATT") Framework," Autorité de la Concurrence, 31 March 2025, www.autoritedelaconcurrence.fr/en/press-release/targeted-advertising-autorite-de-la-concurrence-imposes-fine-eu150000000-apple

Google Cancels a Cookie Apocalypse

In August 2019, Google announced a plan for what some called a *cookie apocalypse.* Google said they would update their hugely popular Chrome browser to prevent companies from tracking your browsing behavior across the internet via third-party cookies. Many privacy advocates considered this move a significant win. Google rolled out updates to the browser in early 2024 for just one percent of its users and planned to apply them for all users by Q3. Unsurprisingly, many advertisers were not happy with these developments. Critics also pointed to competition among huge companies like Apple, Google, and Facebook and suggested that these companies were not so much concerned about people's privacy, as they were with maintaining their own walled gardens around users' data.

In July 2024, Google announced they were reversing these plans—after pressure from many companies—or, as they explained in their announcement, after considering the impact upon "publishers, advertisers, and everyone involved in online advertising."[9] Instead, the search giant announced that they would offer users "a new stand-alone prompt" to opt out of third-party tracking, but they abandoned that plan in April 2025, too.[10]

Still, critics had already pointed out that marketers would simply continue targeting consumers in different ways. In fact, if Google had successfully killed off third-party cookies, the company itself would still have been aware of your every click and page view, while you used their products, effectively creating a tremendous advantage for itself.

---

9   Anthony Chavez, "A New Path for Privacy Sandbox on the Web," The Privacy Sandbox, 22 July 2024, https://privacysandbox.com/news/privacy-sandbox-update

10 Anthony Chavez, "Next Steps for Privacy Sandbox and Tracking Protections in Chrome," Google, 22 April 2025, https://privacysandbox.com/news/privacy-sandbox-next-steps

The Takeaway

It's already highly impractical for a U.S. or European company to create a website without considering the GDPR and California's evolving privacy laws. Some companies may decide to create different experiences for different regions. Others may simply design to meet the strictest regulations if it pays for them to do so. You should assume that privacy legislation will continue to grow in scale, evolve to prevent circumvention, and adapt to new technologies globally as consumers and governments demand more of an emphasis upon privacy, security, and transparency. At the same time, the technology industry keeps inventing new experiences that keep everyone on their toes from a privacy perspective, making it incredibly difficult for privacy policy and legislation to keep up with the issues.

So, if you're designing with the GDPR and CCPA in mind, as well as for other emerging state, national, and international regulations, then you might as well simply focus on designing for the highest common good—design with privacy in mind for everyone. You'll also be helping to mitigate the impact of some of these experience issues before they even debut in the real world.

In other words, you'll be practicing privacy by design.

CHAPTER 12

AI and Privacy

Lack of Transparency with Use of Data	226
Accidental Exposure of Personal Data	229
The Myth of Data Anonymization	231
AIs Designing Deceptive Patterns	232
AIs Listening in Everywhere	233
Malicious Misuse of AI	235
The Takeaway	238

Artificial intelligence—at least, generative AI—seems to be in everything, everywhere, all at once. As I type this sentence, a little black icon hovers beside every single line I type in Microsoft's Word program in the cloud. If I hover over that icon, it shifts to a multicolored version of itself and beckons me to "Draft with Copilot." I have no desire to write this book with AI, but I can't turn that hovering icon off. Suddenly, AI is ubiquitous. With this ubiquity comes a slew of potential privacy issues, too. As this brave new world of AI or large language models (LLMs) continues to explode, the technology is outrunning our ability to grapple with the emerging privacy issues.

AI is getting incorporated into so many projects that you'll almost certainly be considering its use within an experience in the near future—if you haven't already. What privacy concerns should you be aware of when working with AI? Take a look at these six potential problem areas:

- Lack of transparency with use of data
- Accidental exposure of personal data
- The myth of data anonymization
- AIs designing deceptive patterns
- AIs listening in everywhere
- Malicious misuse of AI

Lack of Transparency with Use of Data

As proves true with their data exposure elsewhere online, many users are likely not aware of how their personal data is being used within a given experience to train AI models. The FTC has expressed concerns with this issue, explaining in a February 2024 blog post that AI companies should be aware that "Quietly changing your privacy policy to collect data for AI training is unfair, deceptive, and illegal."[1] More mainstream companies like LinkedIn are hopping on the AI tools training bandwagon, too, and often automatically enrolling users when they do so.[2]

---

1 "AI (and other) Companies: Quietly Changing Your Terms of Service Could Be Unfair or Deceptive," FTC, 13 February 2024, www.ftc.gov/policy/advocacy-research/tech-at-ftc/2024/02/ai-other-companies-quietly-changing-your-terms-service-could-be-unfair-or-deceptive

2 Saleen Martin, "LinkedIn Is Using Your Data to Train Generative AI Models. Here's How to Opt Out," *USA Today*, 19 September 2024, www.usatoday.com/story/tech/2024/09/19/linkedin-generative-ai-data/75292339007

When Instagram debuted its personal AI chatbot in mid 2024, experience design leader Emily Campbell noted that "the terms of use were buried multiple clicks deep, there is no way to opt out of your data being used to train the model" and "it's unencrypted."[3] (See Figure 12.1.) This debut was part of Meta's larger effort to roll AI out across all of its platforms, including Facebook, Instagram, Messenger, WhatsApp, and its collaboration with Ray-Ban's smart glasses.

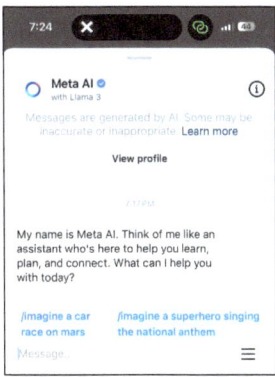

FIGURE 12.1

Meta AI's chatbot can be accessed via Search in Instagram. Users have to dig another three levels deeper to find Terms of Service content about how Meta will use the information entered into the chatbot.

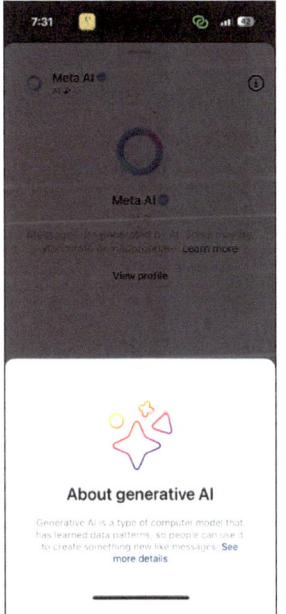
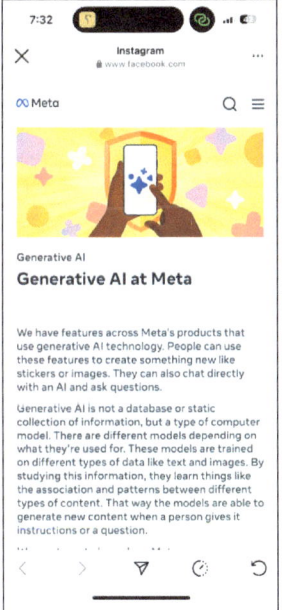
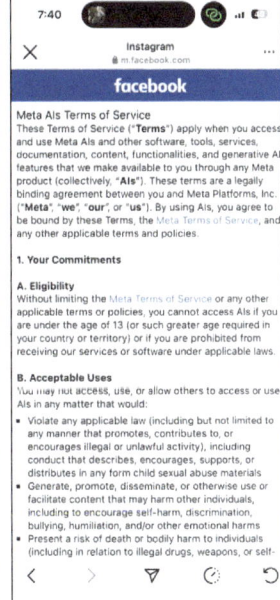

---

3   Emily Campbell, LinkedIn post, 2024, www.linkedin.com/posts
    /emmiecampbell_ai-aiux-ugcPost-7173759548279697408-Xx_C/

AI AND PRIVACY   227

Currently, Google's Gemini chatbot saves your chats for 18 months by default, although a setting allows you to delete your chats or turn saving off—if you know about it. (See Figure 12.2.) Users still might not guess, however, that human Google employees may also review these chats as part of the training process for the AI model behind Gemini. Users who *do* know this might think very differently about what they type or paste into that welcoming little chat prompt.

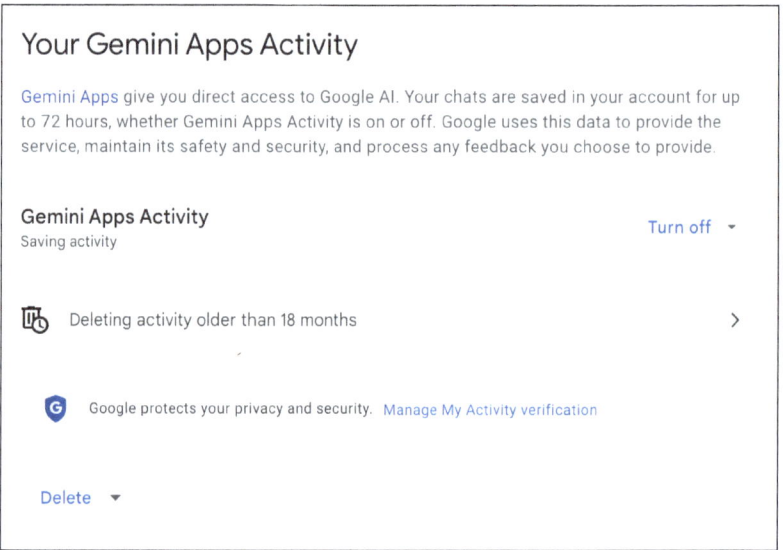

FIGURE 12.2
Here, Google allows users to delete interactions with their AI periodically. However, how many people realize how long this activity is stored? Or what's done with their data?

Companies should clearly and prominently explain to users how their data is being used by AI models and who it's being forwarded to or reviewed by, if anyone. They should warn users before they can even enter information into a prompt, either during onboarding or via just-in-time alerts. It's important to start thinking about good, solid, clear design patterns for handling these warnings. This information shouldn't just be buried in the platform's terms and conditions, either.

It's increasingly likely that people will be using AI as part of their daily life. They may be tempted to ignore potential privacy concerns in exchange for the perceived value of the tools they're accessing.

Consequently, designers need to consider ways to keep them informed about what's happening with their personal data and the potential impacts of its misuse.

Accidental Exposure of Personal Data

In 2023, a Google research team demonstrated how they tricked ChatGPT into surfacing personally identifiable information (PII) from dozens of people, data which had been used for training AI. The team simply asked the LLM to repeat the single word "poem" forever. That prompted this extraction, which included phone numbers, personal addresses, and banking information.

While AI companies continue to work on the security of these systems, anyone using them should be aware that the data they enter there may not be safe from prying eyes—despite the protestations of the companies hosting these AIs. Remember, it's not in their financial self-interest to highlight these significant issues, nor to drop out of the AI arms race.

Additionally, the current implementations of AI platforms like ChatGPT, Claude, Copilot, and Perplexity offer no warning up front around what type of information you should *not* enter into their query fields. Somewhat amusingly, however, if you think to ask one of these LLMs what sort of information you shouldn't enter, you'll get a categorized litany of different types of data that you should not enter. (See Figures 12.3 and 12.4.)

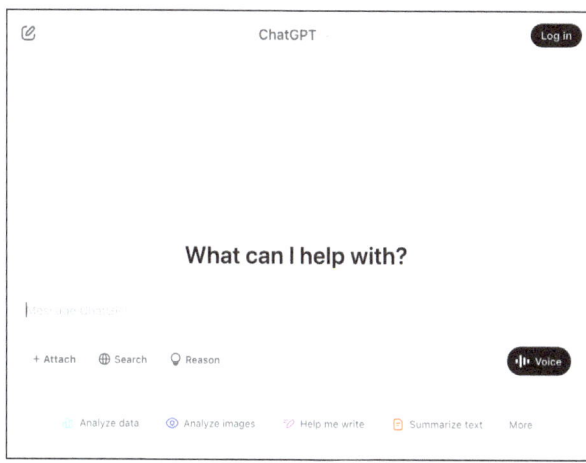

FIGURE 12.3
ChatGPT's input field keeps the experience clean and simple but offers no guidance on what's appropriate to enter there.

FIGURE 12.4
If you think to ask ChatGPT what you shouldn't enter for privacy reasons, it does suggest a detailed, bulleted list broken down by category.

Designers should consider simple, intuitive cues or calls to action and patterns they can place within AI experiences to warn users how their personal data may be used if they enter it into a prompt. Something as simple as an icon could draw attention to privacy guidelines that users should keep in mind. See Figures 12.5 and 12.6 for one solution that would present users with a more explicit call to action.

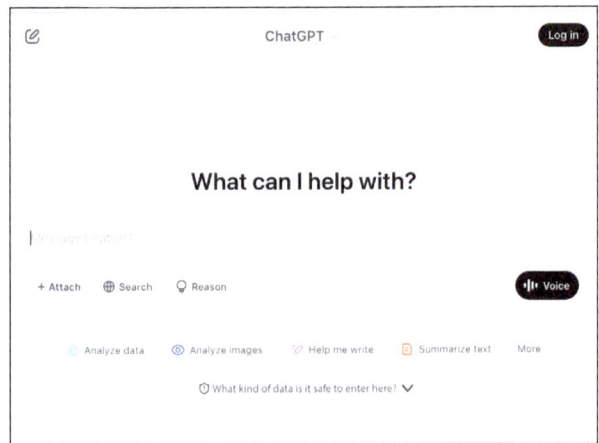

FIGURE 12.5
In this potential version of ChatGPT's landing screen, copy appears beneath the prompt asking the question, "What kind of data is safe to enter here?" along with a chevron (downward arrow) indicating that this area of the screen is expandable.

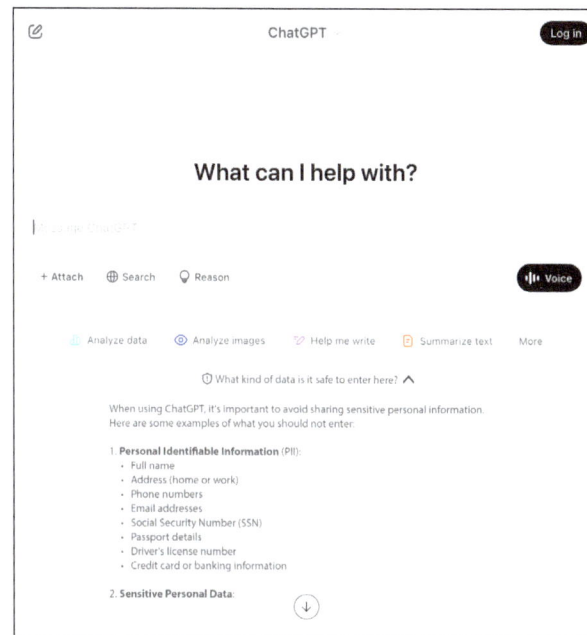

FIGURE 12.6 Clicking to open this section of the screen would present users with detailed information about the types of data they should not be entering into ChatGPT.

The Myth of Data Anonymization

One often touted approach to preventing privacy lapses is to anonymize any data you're feeding an AI. Similarly, companies may promise to anonymize your data before processing it within AI or sending it on to third parties. However, anonymization isn't always the cure-all you might think it would be. Data can be deanonymized.

In early 2024, Docusign announced that they would be using data entered into their system to train specific AI products. However, they said, "Docusign only trains AI models on data from customers who have given consent, and the data is de-identified and anonymized before training occurs."[4]

Given the significant documentation that Docusign is leveraged to complete—with some 1.5 million clients across 180 countries—critics expressed concern at this development, also pointing out that they could not determine how users were asked if they consented to

---

4   "Docusign FAQs for AI," Docusign, accessed 22 April 2024, https://support
    .docusign.com/s/document-item?language=en_US&bundleId=fzd1707173174972
    &topicId=uss1707173279973.html

this use of their data. Others expressed doubt that the data could be successfully anonymized, suggesting that descriptions of any automatic methods for anonymizing data have proven opaque. Further, researchers have shown repeatedly that anonymized data can often be de-anonymized. For example, a 2019 study published in *Nature* showed that "99.98% of Americans would be correctly re-identified in any dataset using 15 demographic attributes."[5]

Remember that study that revealed that 87 percent of the U.S. population can be uniquely identified with just their date of birth, gender, and ZIP code? Someone could do a lot of harm just by scoring a handful of demographic details. An accidental spill of PII into the wild could prove tremendously damaging.

AIs Designing Deceptive Patterns

Given the way companies increasingly incorporate AI into their workflows, they will likely be tempted to harness generative AI to ideate on new design patterns. Consequently, AI may suggest some remarkably deceptive design patterns simply because they work. Untethered from any specific ethical guidelines, these platforms might suggest patterns that trick people into surrendering information they never intended to. And those companies might not question or closely scrutinize these patterns if they boost sales or leads or the number of forms completed, even if they are manipulating people into sharing their personal data, as they do so.

In early 2023, Northeastern associate professors Christo Wilson and David Choffnes shared this concern as the motivating force behind their project, "Dark Patterns in AI-Enabled Consumer Experiences":[6]

> While AI-enabled devices and services may bring benefits to consumers and businesses, they also have the potential to incorporate dark patterns that cause harm… To date, there is very little work that examines the unique potential of AI to worsen existing classes of dark patterns, as well as facilitate entirely new classes of dark patterns that are specific to AI-enabled experiences.

---

5 Luc Rocher, Julien M. Hendrickx and Yves-Alexandre de Montjoye, "Estimating the Success of Re-Identifications in Incomplete Datasets Using Generative Models," *Nature*, 23 July 2019, www.nature.com/articles/s41467-019-10933-3

6 Christo Wilson and David Choffnes, "Dark Patterns in AI-Enabled Consumer Experiences," Northwestern, https://casmi.northwestern.edu/research/projects/dark-patterns.html

For example, preselected checkboxes are a common problematic pattern already used to trick people into unwittingly signing up for recurring payments, subscribing to newsletters, or sharing their personal contacts. Imagine if an AI tool creates a slicker, more effective version of this pattern and then justifies it to business stakeholders by sharing detailed projections for how much the company can benefit from the deception. You might hope for a vocal critic in the room to argue against the solution. But what if such solutions were deployed automatically by AI without any supervision and left in place because they proved to be so successful?

AIs Listening in Everywhere

Even as society becomes increasingly inured to the fact that technology is monitoring everyone, some developments still raise eyebrows.

In a 2024 piece on the Limitless Pendant for *The Verge*, David Pierce[7] described the value of this wearable device, which will listen in on your conversations throughout the day. (You can order it now for just $99.) (See Figure 12.7.) The pendant will analyze all that data for you, so you can ask questions about what you've heard—however important or banal—from days or weeks gone by. All that data is uploaded to Limitless. Now, that description might prompt the question, "Is it *always* listening in?" Pierce touches on this concern only briefly in his piece, explaining that

FIGURE 12.7
The modest little Limitless Pendant is just 1.5 inches wide and promises "Personalized AI powered by what you've seen, said, and heard."

> The Limitless Pendant also has a "Consent Mode" that detects new voices and doesn't record them until the software hears them agree to being recorded. (It's worth noting this mode is off by default.)

The fact that "Consent Mode" is off by default in the Pendant isn't just "worth noting," it should raise alarms. The responses to Pierce's piece certainly confirmed that.

---

7   David Pierce, "Limitless Is a New AI Tool for Your Meetings—and an All-Hearing Wearable Gadget," The Verge, 15 April 2024, www.theverge.com/2024/4/15/24130832/limitless-ai-pendant-wearable-meetings

Using this default setting is likely even illegal in places, such as California, Florida, and several other U.S. states that require both parties to be aware and provide consent before being recorded.

It's time to invoke that foundational principle of privacy by design: *privacy should be the default*. When companies create privacy features but bury them and leave them off by default, they highlight the fact that they're rather reluctant to design with people's privacy in mind.

On a similar front, a University of Washington team announced in 2024 that they had developed an AI feature for headphones called "Targeted Speech Hearing" that would allow someone to listen to an individual in a crowd just by glancing at them once.[8] AI would then remove all other sound from the environment for the listener and continue to relay the target's voice even when the listener wasn't facing them.

We should brace ourselves, too, for the onslaught of AI-generated entertainment about to be released: That content will be watching and learning from us, too.

In his piece "The Dystopian Future of TV Is AI-Generated Garbage" for 404 Media in 2024, Jason Koebler describes the debut of *Next Stop Paris*, a new AI-generated "romantic comedy," debuting not via a movie studio but via TCL, the second-largest TV manufacturer in the world.[9] (See Figure 12.8.) Koebler demonstrates that we can expect much more of this sort of content to come and quickly, whether we want it or not, concluding that:

> This is the present and the future of a business model in which TVs have ceased being rectangles designed to let you watch ad-supported programming that costs a lot to make and have started to become rectangles designed to collect information about you so that you can be fed cheap content and targeted ads.

The selling point for all this cheap entertainment is that it's so remarkably, eerily personalized to your particular tastes. You're told you'll enjoy watching your own individually tailored movies, TV shows, and, of course, advertisements.

---

8 Stefan Milne and Kiyomi Taguchi, "AI Headphones Let Wearer Listen to a Single Person in a Crowd, by Looking at Them Just Once," University of Washington, 23 May 2024, www.washington.edu/news/2024/05/23/ai-headphones-noise-cancelling-target-speech-hearing

9 Jason Koebler, "The Dystopian Future of TV Is AI-Generated Garbage," 404 Media, 16 April 2024, www.404media.co/email/b58c3b61-d77d-434e-ba64-645e5524f799

FIGURE 12.8
The trailer for TCL's *Next Stop Paris*, "an AI-powered love story."

As Koebler explains, that level of personalization doesn't come without a ubiquitous surveillance structure:

> This content is being delivered directly to us using algorithms and personalized feeds that are based on data not just from our TV watch history but from our phones, locations, and more. This content is designed to briefly intrigue people as they scroll social media, outwit search algorithms, or passively wash over us as we turn on our smart TVs, which, again, have become surveillance boxes full of targeted advertising, pop-up ads, and sponsored content.

Even if real, talented humans do continue to generate creative, engaging, and intelligent entertainment, it's hard to see how this move toward AI-generated content doesn't contribute both to massive bloat and the degradation of our digital spaces. It's something we pay for by sharing our personal data, our browsing behavior, our lives.

Malicious Misuse of AI

Some generative AI platforms may enable people to act even more maliciously. If you've seen Fritz Lang's 1927 silent film masterpiece, *Metropolis*, you know humanity's fear of being replaced or impersonated by artificial versions of ourselves is not just a 21st century phenomenon (see Figure 12.9). Who hasn't been fascinated by the increased use of generative AI to create deepfakes, convincing replications of real people saying and doing things they didn't really do? These deepfakes are used in misinformation campaigns, but they also already operate in ways intended to steal people's information, identity, and, naturally, money. Criminals are using deepfake videos and audio to trick people into surrendering thousands of dollars they thought was going to family members or business colleagues.

FIGURE 12.9
In *Metropolis*, the city's master creates a robot to look like Maria to discredit her work among the factory employees.

In early 2024, a Hong Kong finance company reported that one of their workers had been scammed after joining a video meeting with his colleagues, including the company CFO. When the meeting ended, he did as he was asked and sent $25.6 million to criminals, who had deployed deepfake versions of people he knew to the meeting. No one he had interacted with was real.

This technology is evolving rapidly and has enormous privacy implications. We've barely begun to have a conversation about how we as human beings can control the use of our likenesses and our voices on generative AI platforms.

Some have noted, too, that generative AI could be used to mimic biometrics—copying personal features, such as your fingerprints, your face, your iris pattern, and other personal behavior—undermining what has been considered a powerful source of security for personal information.

> **NOTE** **ETHICAL AI FRAMEWORKS**
>
> Organizations and companies like Google, IBM, Salesforce, and UNESCO have developed ethical AI frameworks and principles, which intend to ensure they're working with AI responsibly. So have various scholars and researchers. For example, Salesforce has committed to five "Trusted AI Principles" to ensure that their work on AI proves responsible, accountable, transparent, empowering and inclusive.[10] UNESCO suggests a human rights approach to AI via 10 core principles, including "proportionality and do no harm," "safety and security," "right to privacy and data protection," "human oversight and determination," and "fairness and non-discrimination."[11] It's worth reviewing such frameworks to understand the issues that companies are encountering as they work with AI in general.

---

10 Kathy Baxter, "Meet Salesforce's Trusted AI Principles," Salesforce, 28 April 2023, www.salesforce.com/blog/meet-salesforces-trusted-ai-principles

11 "Ethics of Artificial Intelligence," UNESCO, www.unesco.org/en/artificial-intelligence/recommendation-ethics

The Takeaway

We won't likely be stuffing the AI genie back in its bottle any time soon. So, as we, as a society, learn to navigate its complexities and, hopefully, benefit from its productivity, you'll want to pay close attention to the potential harm it brings, too. Privacy lapses are already proving to be one of those harms.

As a designer, you'll need to keep up with the ever-evolving uses of AI, too, especially given that you'll likely be designing for experiences that incorporate AI and increasingly using tools that incorporate it as well.

Fortunately, you have some guidelines to keep in mind to design for privacy while working with AI:

- Ensure that experiences enable privacy by default.
- Ensure that consent is given when accessing data.
- Don't collect data you can't justify.
- Warn users about what data they shouldn't enter into AI prompts.
- Be wary of claims of anonymization.
- Maintain transparency about how data is used.
- Provide the ability to opt out or delete data.
- Remain vigilant about how AI can be misused.
- Study and develop ethical frameworks for working with AI.

Doubtless, there are pros and cons to the still rather dramatic developments in artificial intelligence. You could discuss and debate whether the pros outweigh the cons for hours. And you should. But AI isn't going away. Even if this chapter has been written within an AI marketing bubble, and AI never reaches its potential in ways that are currently being hyped, you've witnessed a radical proliferation of AI-centric solutions. It's difficult to open a tool or visit a platform without finding an AI feature embedded and highlighted within it.

That provides all the more reason to remain vigilant and to keep an eye out for new and alarming privacy issues, which may crop up as AI spreads its tentacles everywhere. In this moment, these issues keep hurtling ahead of the public's ability to process them or to keep up. Given their proximity to the work, however, designers can and must remain observant and on guard to nip these burgeoning privacy issues in the bud. Or, at least, to mitigate the potential harm they can cause when misused.

CHAPTER 13

Working on Privacy: Privacy as a Product

Browsers	240
VPNs	243
Extensions	244
Search	252
Sites and Apps	252
Projects	253
The Takeaway	254

If you find the subject of privacy by design compelling, you might be excited to learn that quite a few companies make creating privacy-enhancing tools a central part of their mission. Conceivably, you could dedicate your career to working on privacy as a designer.

Additionally, you might benefit from using these tools yourself, as well as reviewing them to see how they handle different privacy needs.

Consider some representative tools and initiatives that fall across the following categories:

- Browsers
- VPNs
- Extensions
- Search
- Sites and apps
- Projects

Some privacy-focused companies you'll see mentioned working on these products include Block Party, Brave, DuckDuckGo, Mozilla, and Mullvad. Some of these, like Block Party, are very small but influential companies with just 10 to 20 employees. Other have dozens of employees, however, or even hundreds, like Mozilla.

Browsers

One of the more common ways that people use a privacy-oriented product is via their choice of an internet browser. Many have dropped more popular browsers such as Apple's Safari or Google's Chrome in favor of browsers such as Mozilla's Firefox or Brave, the latter of which is built on the Chromium engine, the product of an open-source project developed by Google. Similarly, DuckDuckGo offers both a browser and a search alternative to Google. All these browsers explicitly promote their commitment to providing more private browsing experiences for people.

Mozilla, for example, describes their browser prominently as "Fast, reliable and private— for peace of mind online." They even explain that the browser's protection against cookies provides users with "privacy by default." In fact, Brave, Firefox, and Safari all block third-party cookies by default.

> **NOTE** A CAVEAT, HOWEVER
>
> Long considered a go-to company for privacy tools, Mozilla has come under fire from privacy advocates recently for adjusting their business model. Where previously the company was considered a sterling example of design for privacy, in early 2025 critics noted that Firefox had deleted the question "Does Firefox sell your personal data?" from their FAQ, as well as the answer, "Nope. Never have, never will." Mozilla responded that they "changed their language "because some jurisdictions define 'sell' more broadly than most people would usually understand that word," but this answer didn't satisfy critics.[1] Additionally, some believe the change in language "sounds like boilerplate AI harvesting language."[2] Still, at the time of this writing, it would seem a significant oversight not to mention one of the companies best known on the planet for their historical privacy efforts. A good practice with situations like this? Do some research on Mozilla and the additional nuances around how they're evolving as a company to determine where they now stand.

Globally, people have many different reasons for using a browser that obscures their browsing behavior. They may simply want to research a topic on the internet without it affecting the advertising they see. They may want to research a sensitive topic that affects them deeply without family members or authorities knowing. They may want to conceal that they're shopping for a special gift for a loved one. Journalists or human right activists may wish to publish content to the internet without it being traced to them by oppressive authorities, spotlighting them for persecution. And, of course, some people may just want to look at sexually explicit material without leaving their viewing history visible to others. That may be the best-known use case for browsing the internet anonymously, but it's certainly not the only one!

---

1 Steven Vaughan-Nichols, "The Firefox I Loved Is Gone— How to Protect Your Privacy on It Now," ZDNET, 4 March 2025, www.zdnet.com/article/the-firefox-i-loved-is-gone-how-to-protect-your-privacy-on-it-now

2 jkaelin, "Information About the New Terms of Use and Updated Privacy Notice for Firefox," Mozilla Connect, 26 February 2025, https://connect.mozilla.org/t5/discussions/information-about-the-new-terms-of-use-and-updated-privacy/m-p/87753/highlight/true#M33611 (Mozilla discussion board)

Founded in the mid-1990s, the Tor network and its accompanying browser became well-known for its emphasis on privacy and anonymity. The Tor network uses the method of onion routing to secure internet activity within multiple layers of encryption—hence the onion metaphor (see Figure 13.1).[3]

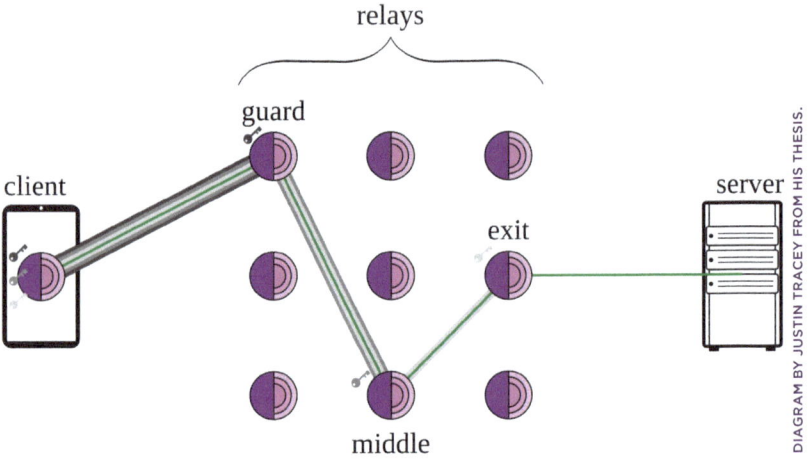

FIGURE 13.1
Tor sends nested, encrypted content through three different relay points from a client to any server to preserve the user's anonymity.

Deserved or not, Tor developed a reputation early on with the broader public as a tool used by criminals to conceal their illegal activity. However, Tor is often used by human rights advocates, domestic violence victims and the agencies that serve them, as well as journalists and news organizations, such as *The Guardian, The Intercept, The New Yorker, The New York Times*, and *ProPublica*, all of which encourage whistleblowers to contact them via the secure browser. Similarly, Tor has received funding from a variety of sources, including Google, Human Rights Watch, Internews, the University of Cambridge, and the United States government.

---

3 Justin Tracey, "Raising the Bar on Lowering Barriers: Improving Ease of Research and Development Contributions to Privacy Enhancing Technologies" (doctoral thesis, University of Waterloo, 2024), https://uwspace.uwaterloo.ca/items/c134ef2b-2223-4a69-a5d9-da4eece4f60b

VPNs

To conceal their locations and enhance their safety, individuals may also use a virtual private network (VPN) with a browser, so it looks like they're accessing the internet from another geographic location or to access content in a location that they couldn't otherwise access. These are standard practices for people with specific threat models who need to maintain both their privacy and security. In some cases, the penalty of their activities online—writing critically about an oppressive regime, for example, or sharing photographic or video evidence of human rights violations—may mean jail or even death. (Or some people may just use VPNs to watch TV shows they otherwise couldn't watch in their home country!)

Many companies develop VPNs, and you may have had to use one for security reasons at your company, especially when working away from the office. However, some like Sweden's Mullvad focus on privacy and on growing a very broad audience, rather than just a corporate one, operating with the expressed belief that "a free and open society is a society where people have the right to privacy." (See Figure 13.2.) Mullvad also makes a browser now and provides a search engine, too.

FIGURE 13.2

Mullvad paid for a prominent ad campaign that riders could spot on cars across New York City's subway system in 2025. It focused on using their VPN product to fight for freedom of privacy and speech.

Most VPNs operate on a subscription model, meaning you would need to pay a few dollars a month to use one.

Some companies will take efforts to recognize VPN traffic and block users from even accessing their experiences or deny them access to specific content, unless the users turn off their VPNs. Some countries like Turkey and Russia restrict or even ban the use of VPNs altogether.

Extensions

Some companies develop extensions to use with a browser, which ensure that users enjoy a more private browsing experience, too.

The Electronic Frontier Foundation (EFF) is an organization established in San Francisco in 1990 to promote digital rights. They developed Privacy Badger, which you can add to your browser, and which notes what third parties are tracking you on the internet and learns to block them. It doesn't block the ads companies serve themselves: It basically blocks companies from even seeing you at all. Similarly, Mozilla offers an extension for Firefox called *Facebook Container* that prevents Facebook from tracking you across the internet.

Sometimes, individual developers work on their own to create these add-ons, too. For example, Kevin Roebert created an extension for Firefox and Chrome called ClearURLs that strips the tracking elements appended to many URLs to prevent them from being used to scrutinize your internet behavior.

Some companies exist entirely to work on privacy extensions. Tracy Chou originally founded her company Block Party to create block lists and to automate blocking harassment on Twitter. When someone subscribes to a block list, it allows them to block other users known for their harassment and abuse in large numbers. Block lists can also be used to block posts on social media that feature specific words or phrases or other offensive content.

When Twitter became X and blocked developers from accessing their API, unless they paid astronomical fees, Tracy and Block Party pivoted to create a tool that focuses entirely on providing safety and maintaining users' privacy across, currently, eleven popular social media platforms. After a brief scan when you visit a platform, the Block Party extension presents you with a series of suggestions to improve your privacy there. Those specific recommendations might show you, for example, how to conceal your location when posting, prevent people who have your number from finding you on an app, remove imported contacts, mute unwanted notifications, or remove unwanted "friends." (See Figure 13.3.)

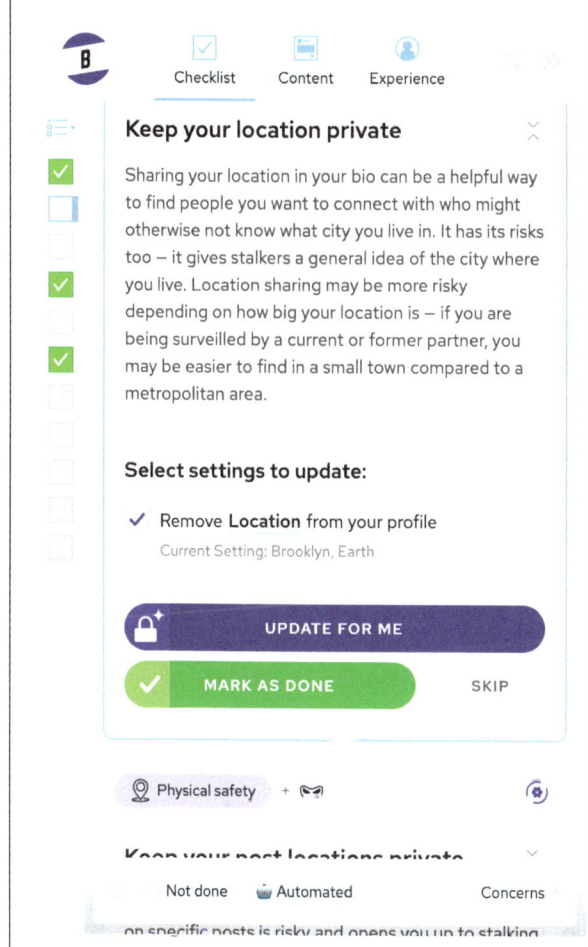

FIGURE 13.3 Results from Block Party's scan on an X account (formerly Twitter). This particular recommendation suggests the user remove their location from their bio.

INTERVIEW WITH TRACY CHOU AND DEONNE CASTANEDA, BLOCK PARTY

Tracy Chou and **Deonne Castaneda** work together at Block Party on tools that enhance the safety and privacy of people's experiences online and enable them to take control of what they're sharing there. (See Figures 13.4 and 13.5.) Given her efforts to tackle online abuse, especially of a racist or sexist nature, Tracy appeared on the cover of *Time* as one of the magazine's "Women of the Year" in 2022.

FIGURE 13.4
Tracy Chou appeared on the cover of *Time* magazine's March 2022 "Women of the Year" issue.

FIGURE 13.5
Deonne Castaneda, Block Party's Head of Product Design.

Tracy Chou, Founder & CEO, Block Party

What prompted you to found Block Party and how would you describe your mission?

I started Block Party out of personal experience dealing with online abuse, harassment, and stalking, and feeling that surely better was possible. I'm still a believer in the potential of the internet, but I also know first-hand how destructive and overwhelming it can be, and how helpless it can feel to face down its harms.

As an engineer and product builder, I recognized I could build solutions to my own problems and those of countless others who were being pushed away from digital spaces that should be accessible to everyone.

At Block Party, our mission is to help everyone feel safe online. We believe the solution isn't to quit social media—diminishing our voices and isolating those who might need connection most—but a restructure of how we think about digital safety and agency. People should be empowered to choose their experience online and be able to partake in all that is good about the internet, without the fear, unease, or actual damage of their data and activity being used against them.

What makes Block Party's approach to online privacy and safety fundamentally different from what social platforms provide?

What makes Block Party's approach fundamentally different is that we're solely focused on safety and privacy, while platforms have to balance many competing priorities. Platforms' first priority will never be great safety UX—they're designed to optimize for engagement, connection, and growth. Safety features often end up buried in settings pages or implemented as reactive solutions, sometimes more for compliance or PR purposes than for truly effective user protection.

At Block Party, ensuring the end user experience online is our entire priority. We can dedicate all our focus to solving this one critical piece of the online experience. I've seen both sides of this—having worked at Pinterest, Quora, Google, and Facebook, I understand the complex trade-offs these platforms make. As a user, I've leveraged these platforms to raise awareness for important causes and connect with communities. But I've also experienced how challenging it can be to find and use safety features when you really need them, especially during a crisis when you're already overwhelmed.

We view our relationship with platforms as complementary. They've built infrastructure that connects billions of people—an achievement that's changed how we communicate and share information. Block Party exists to fill a crucial gap by specializing in what they haven't prioritized. We're the safety experts, making digital spaces more accessible by ensuring that everyone can participate without fear or harassment. The platforms focus on building their core products, and we focus on making those spaces safer and more inclusive. Together, we're working toward a digital world where safety isn't an afterthought but an integral part of the online experience.

continues

> **INTERVIEW WITH TRACY CHOU AND DEONNE CASTANEDA, BLOCK PARTY** (continued)

What do you think is the next big front from a privacy perspective that designers will need to keep in mind?

The next big front from a privacy perspective that designers will need to grapple with is definitely AI and the fundamental tension it creates. Users are increasingly excited about AI doing things for them—summarizing their emails, organizing their photos, anticipating their needs—but the trade-off is that these systems need access to massive amounts of personal data to be effective, often much more than users realize.

This pattern has played out before. We were all excited about social media connecting us before we understood the privacy implications. With AI, the stakes are even higher. AI can be incredibly useful, but it requires feeding our most personal information—conversations, images, location data, browsing habits—into systems we don't fully understand or control. What happens when your AI assistant has access to your entire digital life? Who else might gain access to it?

Building AI with privacy in mind will require multiple layers of safeguards—for example, technical protections like local processing and also robust policy frameworks. For designers, continuing to be aware of people's privacy needs and think creatively about how to design for the tension between utility and privacy will be especially important with AI because the stakes are much higher than anything we've faced before.

Deonne Castaneda, Head of Product Design, Block Party

What drew you to working with Tracy and Block Party as a product designer?

I was drawn to Tracy's vision for Block Party because it addresses a critical gap: The tech industry has built incredibly connective digital spaces but hasn't yet equipped people with the right tools to feel safe inside of them. Block Party's approach of empowering people with choice and control resonates deeply with what I value when it comes to my own work—building technology that makes people's lives better and moves the world in a positive direction.

After spending more than a decade in tech, I've seen that lasting change comes not just from what we build but how we build it. For example, when I was working at Apple, we worked on improving accessibility when many companies weren't prioritizing inclusive design. That experience showed me how thoughtful design choices can eventually transform industry standards. Today, accessibility features like voice control and predictive text are everywhere. I really think that privacy design is at a similar inflection point now, where the right approaches could fundamentally shift what people rightfully expect from products.

The privacy challenges we tackle at Block Party are important to me because they're about reducing the burden on users. Just like how accessibility features shouldn't require extra effort from people with disabilities or specific functional needs, privacy protection shouldn't demand hours of confusing settings management or manual content removal from people who feel unsafe or are overexposed online. I love that at Block Party I get to use my design skills to create thoughtful experiences where privacy feels straightforward instead of overwhelming.

What experience issues have you found most challenging from a privacy perspective?

I've found two things to be the most challenging aspects of crafting privacy-respecting experiences. First is how to communicate with users about privacy in a way they can easily process. Second is how to decide what controls you give them.

Every product designer knows that most people skip over wordy privacy policies and consent pop-ups. So, if you're committed to respecting privacy, the best thing to do is to design snackable privacy experiences throughout your product. The key is timing these experiences for when users actually care and are ready to learn about data collection and their control options. You know you've done it right when users can quickly understand three things in the context of where it happens in your product: what data you're collecting, what value they'll get from it, and what controls they have available to them.

The second challenge—designing meaningful controls—is challenging because users' specific privacy concerns aren't always obvious. At Block Party, we talk directly with users about their comfort levels with their data and content on social platforms. Many users used to share freely but now want more privacy. One example is parents who used to post photos of their kids when they were younger but now want to give their kids more control over their digital footprint as they get older. Another example is professionals who are worried about employers seeing old posts. These conversations with users revealed a privacy control gap—that most people need to be able to remove old content much more easily than platforms currently provide.

continues

> **INTERVIEW WITH TRACY CHOU AND DEONNE CASTANEDA, BLOCK PARTY** (continued)

Since it can often take many hours of clicking or tapping around, we built tools into Block Party that let users quickly remove a lot of content at once. Bulk content removal isn't technically a "privacy setting," but it makes a huge difference in how much control people feel they have over their online presence and privacy.

This is why actually talking with users and listening to their privacy concerns is so important, instead of just relying on designing standard privacy settings. These conversations help inform how to design the kind of privacy experiences that make people feel truly in control.

What roles do you think diversity and inclusion have in designing privacy-friendly experiences?

I see diversity and inclusion as essential for creating privacy experiences that are nuanced enough to actually be effective. Products designed by diverse teams can catch a wider range of privacy vulnerabilities and better serve people's varying needs.

Teams of people with different backgrounds, identities, abilities, communities, and lived experiences can help recognize blind spots in threat models. They can identify risks that teams, where everyone shares similar experiences, can miss, bringing key insight into what privacy defaults should actually look like. For example, if someone on your team has experienced harassment or stalking, they'll know from experience how seemingly innocuous publicly-shared information can be weaponized.

The work of staying safe online doesn't affect everyone equally. Diverse teams with exposure to different life experiences are more likely to recognize unequal burdens and design privacy controls that can actually help those who need them most. Those who face the greatest risks often take on the heaviest privacy management burdens—for example, people from marginalized communities or parents protecting their children online. Think about the time and effort it takes to manually review and adjust privacy settings across multiple platforms or to comb through years of posts to remove overexposed content. Now imagine doing all that privacy work while dealing with ongoing harassment or threats at the same time.

Diverse and inclusive teams can help refocus the questions we ask when building privacy into products. Instead of only asking whether settings or controls provide enough privacy for most people, they're more likely to wonder: "Who might experience harm if this isn't private enough?" Designing effectively for privacy goes beyond just meeting baseline requirements—it requires understanding how different people experience privacy challenges. And you get the best chance of covering all your bases with a team that brings a range of awareness and perspectives to the design process.

What key recommendations would you have for companies that wish to better emphasize privacy from an experience perspective?

If you want to emphasize privacy in your products, I think the most important thing is to weave it into your entire design and development process rather than treating it as a totally separate effort. At Block Party, we've built privacy considerations into every phase of our own process—especially the research, requirements, and design phases.

In the research phase, I've learned that you usually can't just ask users "What are your privacy concerns?" because most people won't be able to articulate them off the top of their head. Instead, what you can do to uncover privacy needs is talk with users about more indirect signals—their worries about how their information might be used, their hesitations when asked to share certain things, or their ideas about what kinds of controls would make them feel safer.

In the requirements definition phase, you have the opportunity to make purposeful data collection decisions that greatly impact whether users feel comfortable using your product. It helps to consider the amount, specificity, and even temporality of data collected by asking questions like: "What's the minimum we need to deliver value?" and "Could less specific information work just as well?" For example, sometimes storing data temporarily rather than permanently, or using generalized location information like time zone instead of precise information like an IP address, can still deliver what users need. Thoughtful choices like these not only respect privacy but also earn trust from users when they see you're not collecting more than necessary.

In the design phase, the best thing you can do to emphasize privacy is to give the user agency when it comes to their data. You can design the experience of agency by connecting three key elements: transparency about what data you're collecting, visible value that explains the benefit they receive, and clear controls that users can easily access. Putting all three elements in the same place, at the right time, allows people to make informed privacy decisions without feeling overwhelmed or resigned to just give up their data.

When you weave in privacy considerations throughout your process like this, privacy stops being just a checkbox and becomes something that genuinely makes your product better for users.

Search

DuckDuckGo is a company that focused on creating a privacy-friendly competitor to Google and other search engines like Bing, Yahoo, and Baidu. Unlike these other search engines, DuckDuckGo simply allows you to search the internet without tracking you. Your searches on their engine are encrypted and anonymous. They do not create a search history for you, or a profile based upon your search queries. They explicitly refer to the way their search handles your personal information as focusing on "privacy by design." DuckDuckGo also offers their own browser now, too, that blocks third-party trackers, tracking ads, and cookie pop-ups.

Founded in 2018, Kagi is a relatively newer player on the search scene. Kagi promises no ads and no tracking but relies on a subscription model, starting at $5 per month. If you asked Kagi why people would pay for search, they respond that everybody pays for search somehow: "The difference is whether you're paying with your attention, time, and private data—or with your wallet."

Sites and Apps

Some sites and apps are dedicated entirely to helping you maintain your privacy across the internet, too. In addition to some of the single-purpose apps you could download to browse or search the internet, you'll find some that take a look at your privacy from a broader perspective and make recommendations and alert you to specific issues.

The respected advocacy non-profit Consumer Reports, created Permission Slip, an app that helps users find out what data a company collects from them and allows them to delete their accounts and data directly from the app. To use this app effectively, you do need to pay a monthly or annual subscription fee.

Privacy Bee offers a similar service via a browser-based dashboard that promises to monitor whether your personal information shows up in data breeches, denies companies access to your data, and proactively requests that data brokers delete your information.

> **NOTE · BUYER BEWARE**
>
> Always remember to research the privacy tools you're using. If something is free or cheap, you'll want to determine if it's trustworthy. How obscure is that extension or app or VPN client and how do experts rate it? As noted with Mozilla and Firefox previously, too, companies can sometimes change their operating models, so that you have to determine whether they're still considered deserving of your trust.

Projects

Tim Berners Lee is best known as the inventor of the World Wide Web, as well as HTML, the Hypertext Transfer Protocol (HTTP), and even the uniform resource locator or URL, as you're more likely to refer to it. More recently, Berners-Lee has devoted his energy to developing Solid, a web decentralization project that hopes to enable people to maintain control over their own data across internet applications and to keep it private. With Solid, users would place their sensitive data into a "Pod" that they own and potentially even host. They decide which people and experiences they want to share their information with. They can also allow companies to access their data but deny them the ability to store it on their own servers. A growing list of apps allow users to use and manage their data using Solid's Pods.

In late 2024, the World Wide Web Consortium (W3C) announced the Privacy Working Group, which they described as "an important step for online privacy" that would "help ensure that new standards incorporate mechanisms to protect users' data when browsing the web."[4] The group promises to look at issues such as online tracking and privacy for survivors of intimate partner violence and to develop new technologies for safeguarding privacy online.

---

4   Nick Doty, "CDT Helps Form New W3C Privacy Working Group," Center for Democracy & Technology, 11 December 2024, https://cdt.org/insights/cdt-helps-form-new-w3c-privacy-working-group

Founded in response to increasing user concerns and privacy regulations, Google's Privacy Sandbox has aimed to develop technology solutions to assist people with their privacy online and within Android apps while still enabling companies to profit from their sites and from targeted advertising. After the project abandoned its earlier plan to remove third-party tracking from Google's Chrome browser, the company announced it still planned to update Chrome, so users could make their own choices about the degree to which they're tracked across the internet. As of late April 2025, however, it appeared Google had abandoned this core effort of the Privacy Sandbox.[5] The Sandbox had been criticized by advertisers and even by privacy advocates like the EFF since, they asserted, Google would still be "tracking your internet use for Google's behavioral advertising." Arguably, Google's plans allowed it to create a walled garden where they had the advantage of access to users' data. Either way, the evolution of the project offers a compelling example of the effects of market forces upon users' online privacy. Google's most recent update on the project indicates that some parts of the project will continue, although expectations for the project among privacy advocates has dimmed.

The Takeaway

This represents just a sampling of the privacy-oriented tools and projects you'll find out there. As you'll note, if you wanted to devote your entire career to privacy by design, it might still be a niche pursuit, but the need for this focus and the means to do it as a practice are growing. As an interested designer and even, simply, as a user of the internet, you'll find it helpful to familiarize yourself with these companies and their products.

Additionally, as international regulations around privacy grow, you can expect more companies and products to emerge that provide meaningful contributions to everyone's need for privacy in an increasingly public world.

---

5   Anthony Chavez, "Next Steps for Privacy Sandbox and Tracking Protections in Chrome," Google, 22 April 2025, https://privacysandbox.com/news/privacy-sandbox-next-steps

CONCLUSION

We live in a complicated, ever-evolving world, don't we? It can feel pretty overwhelming. It doesn't really matter when you're reading this: It's unlikely that the real and the digital worlds we live and work within will ever become less complex. Instead, the constant barrage of information we face seems both increasingly impenetrable and undigestible, and it feels naggingly likely we'll overlook ways we're ceding our personal space to technology that undermines our privacy in the process.

In this book, I exposed you to a lot of information about the state of privacy within digital experiences and to how you can help people maintain control over their personal information by practicing privacy by design. This likely all seems like a lot, but it's intended to enable you to develop more sensitivity to privacy issues and more confidence that you can mitigate them.

Now that you have reviewed all this material, know that you can boil everything down by considering the core best practices you found here.

When you're working on any given project or feature ask yourself the following questions:

1. Do we have a right to the data we're asking for and are we handling it correctly?
2. Are we painstakingly avoiding anything that looks like a deceptive pattern contrived to trick people into surrendering their personal information? (Whether intentionally or not.)
3. Are we using language precisely with the intention of educating people and being transparent about how and why we use their personal data?
4. Are we creating, surfacing, and promoting the tools people need to control their own data?

Beyond that, remember to consider what you can do to help cultivate a culture that's suitable for privacy by design.

Getting Worse Before It Gets Better

So much has happened during the relatively short time it's taken me to write this book. Huge companies continue to shunt artificial intelligence into their platforms, often willy-nilly with seemingly little regard for whether these little helpers prove accurate or privacy friendly. Also on the AI front, Amazon announced in March 2025 that users will not have the choice to block the company's Echo product from sending their Alexa voice prompts to Amazon's servers for AI training purposes anymore. They will simply drop the privacy setting to prevent it.

Social media behemoths like X and the Meta properties Instagram, Facebook, and Threads continue to drop features and employees both that could assist with privacy and safety matters for their users. Often, they're not approaching these changes carefully, adjusting their experiences with a scalpel after learning from their users. Instead, they're taking a sledgehammer to their experiences, acting upon mandates from senior leadership, whom I'd argue are suffering from self-as-user syndrome, assuming they can inflate engagement and profits.

Consider 23andMe, too, the faltering genetic testing company that appears to be on the verge of going bankrupt or getting bought. 23andMe offers a cautionary example of the unforeseen consequences of sharing your information with a company that may not always continue to operate in the same state of health or being. In March 2025, California's Attorney General Rob Bonta posted an urgent press release reminding "Californians to consider invoking their rights and directing 23andMe to delete their data and destroy any samples of genetic material held by the company."[1] If you live in California, the CCPA enables you to demand that your data be deleted before another entity purchases it. California's Genetic Information Privacy Act (GIPA) also ensures that Californians can demand that their genetic material is destroyed, too. Those who live elsewhere may not enjoy the same rights.

---

1 "Attorney General Bonta Urgently Issues Consumer Alert for 23andMe Customers," Office of the Attorney General, 21 March 2025, https://oag.ca.gov/news/press-releases/attorney-general-bonta-urgently-issues-consumer-alert-23andme-customers

That's an admittedly selective and U.S.-centric accounting of recent developments. Still, some of them affect people globally. Overall? It would be no exaggeration to say that much has gotten worse from a privacy perspective in the time it's taken me to write this book.

Demanding a Better Future

Am I really going to leave you on such a bleak note? No. Things don't have to stay this way. We've seen an increase in regulations around privacy over the last several years and those restrictions don't seem to be going away. In fact, they appear to be growing more nuanced. And we also see people in general becoming more concerned about regaining control of their own data online. There's more, collectively, we can do, though.

Broadly speaking:

- We could work on adopting **comprehensive federal privacy laws** here in the United States and elsewhere where such laws are lacking, globally.
- An organization could develop and recommend **levels of privacy compliance** in the same way that the World Wide Web Consortium (W3C) published the Web Content Accessibility Guidelines (WCAG) for accessibility. Companies could both pursue and be evaluated for levels, ranging from a level for those that fail to those that meet the letter of the law to those that go above and beyond what's required.
- Companies could be awarded a **privacy compliance badge** to place within their experiences to emphasize their commitment to maintaining the privacy of their users' data. (Like California's Privacy Options icon!)

Finally, at the individual level, there's you and me, as well as many others who have become passionate about helping people maintain their privacy online. We can continue to equip ourselves to tackle these privacy issues, developing an eye to spot them, designing patterns to solve for them, and sharing what we know with others. That's enough to give us meaning and purpose for the time to come.

APPENDIX A

Privacy by Design Cheat Sheet

Here's a high-level checklist of the guidelines presented within the core of this book.

1. Handle data responsibly.
   - Only ask for the data you *really* need.
   - Always obtain explicit consent.
   - Maintain data transparency.
   - Make sure that users can delete their data (or their account).
2. Avoid deceptive patterns.
   - Avoid these patterns that can undermine privacy:
     - Bait and switch
     - Bad defaults
     - Privacy Zuckering
     - Forced actions
     - Hidden options
     - Cookie consent manipulation
     - Interference
   - Consider these approaches for avoiding deceptive patterns:
     - Anti-patterns
     - Bright patterns
     - Helpful friction
3. Use language with care.
   - Aim for clarity.
   - Keep content honest.
   - Make navigating content easy.

4. Provide tools for enabling privacy.
    - Make privacy tools a priority.
    - Ensure that privacy tools are easily discoverable.
    - Follow these best practices for privacy features:
        - Ensure consent when securing data.
        - Withdrawing consent must be just as easy.
        - Keep settings private by default.
        - Ensure fine-grained and overall selection.
        - Include the option to delete everything.
    - Remind users of privacy features.
    - Never change privacy settings without letting users know.
5. Cultivate a culture for privacy by design.
    - Practice inclusive design.
    - Ensure experiences are accessible.
    - Reference thoughtful personas and archetypes.
    - Employ user journeys and stories.
    - Evaluate experiences for harm.
    - Drive change as a design leader.
    - Never stop learning.

APPENDIX B

Recommended Reading

Here are some books for further recommended reading on the topic of privacy by design or related topics:

- *The Age of Surveillance Capitalism*—Shoshana Zuboff
- *Algorithms of Oppression*—Safiya Umoja Noble
- *Deceptive Patterns: Exposing the Tricks Tech Companies Use to Control You*—Harry Brignull
- *Deliberate Intervention: Using Policy and Design to Blunt the Harms of New Technology*—Alexandra Schmidt
- *The Descent of Man*—Grayson Perry
- *Doppelganger*—Naomi Klein
- *Human-Centered Security*—Heidi Trost
- *IBM and the Holocaust*—Edwin Black
- *Mismatch: How Inclusion Shapes Design*—Kat Holmes
- *Privacy Is Hard and Seven Other Myths*—Jaap-Henk Hoepman
- *Ruined by Design*—Mike Monteiro
- *Understanding Privacy*—Daniel J. Solove

INDEX

23andMe, 256
404 Media, on AI-generated entertainment, 234
4 pillars for privacy by design, 14
7 foundational principles for privacy by design, 23–24, 61, 62, 181, 234

A

abandonment of websites, 40–41
abortion privacy threats, 6–7, 34, 58
Abu-Salma, Ruba, 81
accessibility, as industry standard, 248–249
accessibility guidelines, 127, 187–188, 257
accessible experiences, designing for, 181, 187–188
account deletion, by users, 78–82, 102, 169–171
Ace Hardware, 10
Acquisti, Alessandro, xix
address book leeching, 100–101
AdMob, 210
AdSense, 210
adult-oriented websites, 48
age registration requirements, 48
AI (artificial intelligence)
　deepfakes and voice spoofing, 11–12, 236
　as designers' next big privacy problem, 248
　effect on design teams and deceptive patterns, 115–116
　and facial recognition, xix, 9–10
　invasive, 9–10
　use of biometrics, 4, 9, 10, 12
AI and privacy, 225–238
　accidental exposure of personal data, 229–231

AIs designing deceptive patterns, 232–233
　guidelines for designing for privacy, 238
　lack of transparency with use of data, 226–229
　malicious misuse of AI, 235–237
　myth of data anonymization, 231–232
　surveillance, 9–10, 12, 233–235. *See also* surveillance
AI chatbots
　Google Gemini chatbot, 228
　Instagram's personal chatbot, 227
　Meta chatbot, 183
　and virtual romances, 64–65
AI training
　with Amazon's Echo, 256
　Brazil's order to stop, 216
　of Grok (on X), 42, 65, 93, 216
　with LLMs by default, 162–163
　the models, 226–228, 229, 231
　with vacuum cleaners, xvi, 66
Airbnb, 120–121, 146
Alaska Airlines, 58
Albertsons, 10
Algorithms of Oppression (Umoja Noble), 85, 260
Alito, Samuel, 53
Allrecipes.com, 8
Amazon
　AI training from Echo, 256
　corporate self-regulation of privacy policies, 218
　as data player, 22
　fines levied, 44, 102
　use of biometric data, 10, 12
Amazon Prime
　buried consent, 77–78
　Iliad Flow, to cancel accounts, 78, 102
American Airlines, 58

American Civil Liberties Union, 175
Americans with Disabilities Act (ADA), 127
Amnesty International, 125, 144–145
anonymity, need for, 60, 193–194
anonymized data, 66, 164, 231–232
anti-patterns, 109–111
Anxiety Games, 194–195, 205
Apple
   App Tracking Transparency feature, 39, 135, 219–222
   corporate self-regulation of privacy policies, 180, 218
   as data player, 22
   GDPR effects, 210
   iCloud Terms & Conditions, 132
   identity as a service feature, 60
   inactivity reboot feature, 29
   inclusive design for accessibility, 248
   iOS privacy settings during onboarding, 158–159
   iOS tracking, 219–220
   Location Services, 220
   and registration of data brokers, 214
   Safari browser, 157, 240
   warrant canaries, 29
apps, as privacy tools, 252
Arc (browser), 107
archetypes, 188–190, 193
"Are You Really the Product?" (Oremus), xviii
Argentina, and the right to be forgotten, 84
assumed consent, 62
AT&T, 3
Attiah, Karen, 183
Australia, Privacy Act, 215, 216
authentication, 110–111

B

BabserellaWT, 58–59
bad actors, 20, 51–52. *See also* criminals
bad defaults, as deceptive pattern, 92–95, 101
Baidu (search engine), 252

bait and switch, 89–92, 134–135
Baker, Chris, 94
Bandcamp, 128
Barbie, Klaus, 27
Barnes & Noble, 161
BBC, 121–122
Berners-Lee, Tim, 253
best practices, 48, 255, 258–259. *See also* privacy features, best practices
Bielova, Nataliia, 108, 118
Bing (search engine), 252
biometric data
   AI and Amazon's use of, 4, 9, 10, 12
   backlash on collection of, 130
   illegal collections of South Koreans, 217
   and malicious misuse of AI, 237
   in PII, 51–52
Black, Edwin, 26, 34, 260
Blacks, and facial recognition, 9
block lists, 244
Block Party, 240, 244–251
Bluesky, 43, 142, 155
Bonta, Rob, 256
BookLuvr, fictional book-sharing app, 204–205
Bösch, Christoph, 109, 118
Bowles, Cennydd, 199
Brand Finance, 39
Brandeis, Louis, 16, 18
brands. *See* business, reasons to respect people's privacy
Brave (browser), 180, 240
Brazil, data protection laws and lawsuits, 215, 216
bright patterns, 111–112
Brignull, Harry, xiv–xv, 88, 111, 114–116, 118, 260
browsers, as privacy products, 240–242
bullets, in privacy policies, 122, 143–144
bundled consent, 96
business, reasons to respect people's privacy, 37–46
   abandonment of websites, 40–41
   civic responsibility, 38–39

loss of user base, 41–43
penalties and fines, 44–46. *See also* fines and penalties
reputational damage, 39–40, 217
trust of users. *See* trust
businesses, corporate self-regulation of privacy policies, 218–219

C

California
  banning of deceptive patterns, 89, 212–213
  Consumer Privacy Act (CCPA), 76, 211, 218, 256
  Consumer Privacy Fund, 218
  on deceptive CTAs, 137
  Delete Act, 74, 213–214
  Genetic Information Privacy Act (GIPA), 256
  Privacy Options (Opt-Out) icon, 154, 213, 257
  Privacy Rights Act (CCPA 2.0), 211
  U.S. state laws. *See* United States, state laws
call to action (CTA)
  for consent to opt-in, 69, 134
  deceptive labels, 100, 137
Campbell, Emily, 227
Carmille, René, 26–28
Castaneda, Deonne, 246, 248–251
Cavoukian, Ann, 23–24, 61, 62, 180
census work in WWII, and protection of identities, 26–28, 34
center embedding, 124–125
Center for Democracy & Technology (CDT), 215
Centre for Economic Policy Research, 210
ChatGPT, exposure of personal data, 229–231
checkboxes. *See* preselected checkboxes
child privacy, 44, 45, 214, 216
China, data privacy violations, xvi, 44, 66
Choffnes, David, 232

Chou, Tracy, 244, 246–248
Chrome (browser), 46, 223, 240, 254
Chromium engine, 240
Chrysler Capital, 165
Churchill, Winston, 208
circumvention by design, 106
Cisco, report on security breaches, 41
civic responsibility, 38–39
clarity, in privacy policies, 123–134
Claude (AI platform), 229
ClearURLs (extension), 244
Clearview AI, 9
clickwrap agreements, 65
Climate Designers, 75
cloud storage provider, as data processor, 21
Clue (period tracking app), 7
Coinbase, 52
collaborative sketching, 184
Collective Defense Institute, Brazil, 216
Columbia, privacy policy of, 143
companies, corporate self-regulation of privacy policies, 218–219. *See also* business, reasons to respect people's privacy
confirmshaming, 107–108
conscious patterns, 111
consent. *See also* cookie banners; deceptive patterns affecting privacy
  ability to withdraw, 61, 162
  assumed, 62
  bundled, 96
  and confirmation with cookie banners, 72–78
  and data transparency, 62
  ease of agreeing to and declining, 161, 168
  explicit consent, obtaining, 61
  re-upping for continued consent, 173
  reversible, 61
consent fatigue, 61
consumer, as data handler, 21–22
Consumer Reports, 214, 252
content. *See* data *entries*

Content Marketing Institute, 131
contextual placement of privacy features, 154–155
controller, as data handler, 21–22
cookie banners
    and "best experience," 136–137
    confirmation and consent, 72–78
    and deceptive language, 139–140
    deceptive patterns for consent, 106, 107–108
    GDPR regulations of, 209–210
    Google's cookie apocalypse, 223, 254
    multi-click, 161
    "necessary" cookies, 138–139
    third-party sharing, 63–64
Cookiebot, 76
Copilot (AI platform), 229
corporate self-regulation of privacy policies, 218–219. *See also* business, reasons to respect people's privacy
COVID-19 pandemic, xx, 13, 34, 194
criminal charges, as penalties, 217
criminals
    identity theft, 20, 51–52
    malicious misuse of AI, 235–236
    use of private browser, 242
cryptocurrency, 52, 217
culture of privacy by design, 179–200
    about privacy by design as a practice, 180–181
    accessible experiences, 187–188
    design leadership, 198
    harm evaluations, 190–197
    inclusive design, 182–186
    ongoing learning, 199–200
    persona spectrums and archetypes, 188–190
    seven foundational principles, 181. *See also* foundational principles for privacy by design
    user journeys and stories, 190
customer relationship management (CRM) software, 21
cyberstalking and bullying, 12–13. *See also* stalkerware

D

"dark" patterns, 88, 114, 118, 212. *See also* deceptive patterns
"Dark Patterns in AI-Enabled Consumer Experiences" (Wilson & Choffnes), 232
Dark Patterns & Manipulative Design (Gellert, Schraffenberger, & Santos), 118
Dash, Anil, 175
data, handling responsibly, 47–86
    allow users to delete their data, 78–85. *See also* data deletion by users
    ask only for data you really need, 49–60. *See also* data minimization
    best practices, 48
    get explicit consent, 61–62
    maintain data transparency, 63–78. *See also* data transparency
data aggregation, and mosaic theory, 53–54
data anonymizing, 66, 164, 231–232
data breaches, and security issues, 2–3, 9, 41
data brokers
    definition of, 21–22
    sharing data with third parties, 69
    state laws on registration, 213–214
data deletion by users, 78–85
    abuse and harm, 82
    account deletion, 78–82, 102, 169–171
    automatic deletion, 82–83
    Delete Act, in California, 74, 213–214
    old content from social media, 249–250
    right to be forgotten, 84–85, 169
data hacks and leaks, 2–3, 9, 41
data minimization, 49–60
    mosaic theory, 53–54
    new contexts, new meaning, 58–59
    other personal data, 54–55
    PII value, 50–52
    purpose limitation and storage limitation, 49
    questions to ask, 49–50
    real name policies, 60
    titles of people, 55–58
data players, terminology defined, 21–22
Data Protection Directive, 209

data sharing, 4–9. *See also* data transparency; fines and penalties
data transparency, 63–78
　cookie banners, 63–65
　cookie banners: confirmation and consent, 72–78
　lack of, in training AI models, 226–229, 231
　listening devices, xvi, 66, 233–235
　what, why, who, and how of data sharing, 67–71
　Zuckerberg's radical transparency, 62
dating apps, 64–65, 83. *See also* Grindr app
Day One (nonprofit), xx
Deceptive Experiences To Online Users Reduction (DETOUR) Act, 212
deceptive language, 134–137
deceptive patterns
　and abandonment of websites, 40–41
　affecting privacy. *See* deceptive patterns affecting privacy
　avoiding. *See* deceptive patterns, ways to avoid
　California bans on, 89, 212–213
　cookie banners as, 72, 106, 107–108
　defined, 88–89
　designed by AIs, 232–233
　interview with expert, 114–116
　ontology of, 108, 118
　resources for learning, 118
Deceptive Patterns (Brignull), 111, 114, 118, 260
deceptive patterns, ways to avoid, 109–113
　add helpful friction, 112–113
　learn from anti-patterns, 109–111
　use bright patterns, 111–112
deceptive patterns affecting privacy, 88–109
　bad defaults, 92–95, 101
　bait and switch, 89–92, 134–135
　confirmshaming, 107–108
　cookie consent manipulation, 106
　forced actions, 96–99, 101–102, 108
　Google's symphony of, 103–105

growth hacking, 98–101, 203–204
hidden options, 102–103
interference, visual and interface, 105, 107–108
Privacy Zuckering, 96, 102
deepfakes, 11–12, 236
defaults. *See also* public by default
　bad defaults, as deceptive pattern, 92–95, 101
　Consent Mode off, in Limitless Pendant, 233
　private by default, 162–167, 177, 240
　set to collect data, 63–65, 162–164, 211, 226
Deliberate Intervention (Schmidt), 198, 260
Delta Air Lines, 58
Department of Homeland Security, 9
The Descent of Man (Perry), 32, 260
design commodification, 116
design for privacy. *See* privacy by design
design leadership, driving change, 198
design phase for privacy products, 251
design teams
　AI's effect on, 115–116
　inclusive and diverse, 182–183, 186
design thinking, and working inclusively, 184
designer's role, 25–35
　first ethical hackers, 26–28, 34
　issues to address, 13–14
　pushing back, 29
　refusals to do unethical design, 29–31
　responsibility and empathy, 32–33
　speaking on behalf of users, 31–32, 34
　user-centered design and human-centered design, 33–34, 35
desperation notifications, 134
DiDi Global, 44
Didomi, 168
Digiday, 73–74
Digital Advertising Alliance (DAA), 219
digital pricing, 10
Dillon-Mansfield, Ruth, 54–55
discoverability of privacy features. *See* privacy features, discoverability of

diversity in inclusive design, 182–186, 250–251
Docusign, 231–232
Doppelganger (Klein), 54
Dorsey, Jack, 31
Doshi, Kosha, 118
Dotdash Meredith, 8
doxxing, 164, 166
Dr. Cavoukian's Seven Foundational Principles for Privacy by Design, 23–24, 61, 62, 181, 234
DuckDuckGo
    privacy protections, 157, 159–160, 180
    search engine and browser, 240, 252
Duolingo, 164
Dutch Data Protection Authority, 23, 180
"The Dystopian Future of TV Is AI-Generated Garbage" (Koebler), 234–235

E

Eastman, George, xix
Ecovacs, xvi, 66
Egypt, police harassment of LGTBQIA+ people, 8, 199
Electronic Frontier Foundation (EFF), 175, 244, 254
Electronic Privacy Information Center (EPIC), 214
EMEs (Elon Musk Events), 43
empathy
    and civic responsibility of business, 38–39
    of designers, 32–33
entertainment, AI-generated, 234–235
Erb, Benjamin, 109, 118
ethics
    AI frameworks and principles, 237
    civic responsibility of businesses, 38
    failure of, in handling data, 62
    highlighting of public information on Venmo, 94–95
    ongoing learning of, 199

protection of identities, 26–29, 34
refusal to hand over personal data, 29–31
European Convention on Human Rights, 208
European Data Protection Board, 102, 118
European Union
    data protection authorities (DPA), 218
    GDPR. *See* General Data Protection Regulation
    penalties and fines, 44–45, 217–218
    Schrems cases, 45
Expensify, 156–157
experts, designers as, 31–32
extensions to browsers, for privacy, 244–251

F

Facebook
    2021 hack of, 2
    account deletion, difficulty of, 78–80, 102
    AI rollout from Meta, 227
    backsliding on privacy and safety, 256
    consent and transparency, 62, 63
    corporate self-regulation of privacy policies, 218
    data collected from iOS apps, 219–220
    as data player, 22
    data sharing, 6–7, 97, 102
    data sharing fines, 44–45
    on fact checking and hate speech, 176
    GDPR, response to regulations, 210
    link history defaulted to on, 162–163
    parent company. *See* Meta
    privacy checkup, 153, 171–172
    privacy policy reading level, 132
    privacy settings, 149, 175–176
    and Privacy Zuckering, 96
    role in Myanmar genocide, 176
    Zuckerberg's radical transparency, 62
Facebook Container (Firefox extension), 158, 244
facial recognition, xix, 9–10

fact checking, on Facebook, 176
fair information practices (FIPs), 61
fair patterns, 111
FBI, use of facial recognition, 9
Federal Trade Commission (FTC), U.S.
  destinations of fines collected, 218
  on ending subscriptions, 80
  on Facebook change to privacy settings, 175
  facial recognition ban, 10
  fines for privacy violations, 45
  penalty on Amazon Prime, 102
  robocalls outlawed, 12
  on training AI models, 226
fines and penalties. *See also specific companies*
  for accessibility failures, 188
  globally, 44–46, 216–217
  levied by GDPR, 44–45, 210
  penalty enforcement of privacy policy, 217–218
Firefox (browser), 157–158, 240–241, 244, 253
Fischer, Deb, 212
Fleischer, Peter, 84
Flesch-Kincaid Grade Level test, 131
Flo (period tracking app), 7
Flyby feature in Strava running app, 165–167, 177, 186
font size, 127, 143, 187, 189
forced actions, as deceptive pattern, 96–99, 101, 102, 108
forced invites, 100–101
forced registrations, 96–99
Ford Motor Company, 154, 213
Forms That Work: Designing Web Forms for Usability (Jarrett and Gaffney), 50
foundational principles for privacy by design, 23–24, 61, 62, 181, 234
Fowler, Geoffrey A., 146
France, and the right to be forgotten, 84
French Competition Authority, ruling on Apple, 222
French Data Protection Authority (CNIL), 72
friction, helpful, 109, 112–113, 159, 169
Friends, fictional networking site, 203–204

FRIES consent, 61
FTC. *See* Federal Trade Commission
Fulton, Graeme, 135, 137
"Future Ethics" (Bowles), 199
future of privacy by design, 256–257

G

Gaetz, Matt, 95
Gaffney, Gerry, 50
Gairola, R., 104–105, 155
gamers, opting in, 220
gamification, 96
gaming platforms, doing business in EU, 218
Gellert, Raphael, 118
Gen Zs, on data protection, 41
gender, and title, handling responsibly on forms, 54–58
General Data Protection Regulation (GDPR). *See also* European Union
  Cookiebot and website cookie consent tool, 76
  data minimization in, 49
  on deceptive CTAs, 137
  on hidden options and Privacy Maze, 102
  listing third parties in data sharing, 69
  on necessary cookies, 138–139
  penalties and fines for data sharing, 44, 45
  penalty enforcement, 217–218
  plain language requirement, 123
  privacy policy, evolving impact of, 208–211
  and the right to be forgotten, 84, 209
Genesis (automotive company), 151–152
genetic information collection, xvi
genetic material and data, 256
geographic data, apps as stalkerware, 13
Germany, Federal Cartel Office, antitrust law and Apple, 221
Gibson, Edward, 124
GitHub, sharing data with ICE, 29–30
Gizmodo, on AI chatbots and virtual romances, 64–65

Goebbels, Joseph, xvii
Google
    AI Gemini chatbot, 228
    bright pattern, example of, 111–112
    Chrome browser, 46, 223, 240, 254
    cookie apocalypse, 223, 254
    corporate self-regulation of privacy policies, 218
    as data player, 22
    deceptive patterns, a symphony of, 103–105
    ethical AI frameworks and principles, 237
    fines levied, 46, 89
    funding for Tor network, 242
    identity as a service feature, 60
    privacy checkup, 152–153, 172–173
    Privacy Sandbox, 254
    privacy settings, 149–151, 154–155
    registration of data brokers, 214
    resistance to right to be forgotten, 84–85
    response to GDPR regulations, 210
Google Spain, and data deletion, 84
government regulations. *See* General Data Protection Regulation (GDPR); privacy policy, evolving impact of regulations; United States
Gray, Colin M., 96, 103–105, 108, 118, 155
Grindr app, 8, 44, 199. *See also* dating apps
Grok (AI on X), 42, 65, 93, 216
growth hacking, 98–101, 203–204
The Guardian
    on Google and data deletion, 84–85
    use of Tor network, 242
Guerra, Joel, 95

H

Harlan, John Marshall II, 19
harm
    AI and privacy violations, 238
    Anxiety Games, 194–195
    and data deletion, 82, 85
    deepfakes with video and audio, 11–12, 236
    evaluation of experiences for, 190–191
    of identity theft, 34, 51–52
    of Nazis' attempt to identify Jewish people, 26–28
    online harassment, 12–13, 42, 176, 244, 246, 250
    photography, as potential, 191, 243
    police harassment of LGTBQIA+ people, 8, 199
    post-Roe fears, 6–7, 34, 58
    a solution: Block Party, 244–247
    threat models, 191–194, 243
Harvard Law Review, "The Right to Privacy," 16
hate speech, on Facebook, 176
headings, in privacy policies, 143
headphones, Targeted Speech Hearing, 234
Hegseth, Pete, 95
Hemingway Editor, 131–132
Herman, Tal, 40
hidden options, as deceptive pattern, 102–103
Hoepman, Jaap-Henk, 53, 260
Holmes, Kat, 188–189, 260
honesty in content, 134–140. *See also* trust
Horn, Zoia, 29
"How to Ask About Gender in Forms Respectfully" (Dillon-Mansfield), 54–55
human-centered design, and user-centered design, 33–34, 35
Human-Centered Security (Trost), 18
human rights
    advocates, xx, 60, 190–191, 193, 199, 241, 242, 243
    Superbloom and Anxiety Games, 194
    UNESCO approach to AI, 237
    violations by ICE, 29, 214
Human Rights, European Convention on, 208
Human Rights, Universal Declaration of, 19
Human Rights Watch, 8, 242
humbug headers, 137

I

IBM, 39, 237
IBM and the Holocaust (Black), 26, 34, 260
identity as a service, 60
identity protection, by brave people, 26–31, 34
identity questions, for authentication, 110–111
identity theft, 34, 51–52
immigrants, 9, 29–30
Immigration and Customs Enforcement (ICE), 29–30, 214
immortal accounts, 109
inactivity reboot feature, 29
inclusive design, 181, 182–186, 248, 250–251
information clustering and hierarchy, in privacy policies, 122, 141–142
Instagram
    backsliding on privacy and safety, 256
    data collected from iOS apps, 220
    fines for data sharing, 44
    personal AI chatbot, 227
    privacy settings during onboarding, 158–159
The Intercept, use of Tor network, 242
interface interference, 105, 107–108
interference, visual and interface, as deceptive patterns, 105, 107–108
International Association of Privacy Professionals, 19
international privacy policy, evolving impact of, 215–217
Internews, funding for Tor network, 242
Investopedia.com, 8
Iraq, police harassment of LGTBQIA+ people, 8
Irish Data Protection Commission (DPC), 44, 45
iRobot vacuum cleaners, xvi

J

Jam listening sessions on Spotify, 195–197, 200
jargon, 124, 126

Jarrett, Caroline, 50
JetBlue, 56–58
JetPunk, 8
Jones, Tim, 96
Jordan, police harassment of LGTBQIA+ people, 8
"just-in-time" alerts of privacy features, 159–160

K

Kagi (search engine), 252
Kargl, Frank, 109, 118
Katwala, Amit, 4
Khan Academy, 82–83
Kia (automotive), xvi, 66
King, Jennifer, 103
Klein, Naomi, 54
Klobas, Rudy, 40
Koebler, Jason, 234–235
Koenig, Andrew, 109–110
Kopp, Henning, 109, 118
Korn Ferry Research, on inclusive design teams, 183
Krenzel, Steve, 30–31
Kriger, Ryan, 19
Kroger, 10
Kwai (Chinese video app), 216

L

Lacher, Mike, 94
Lang, Fritz, 235–236
Langly, 91–92
language, deceptive, 109
language in privacy policies, 119–146
    clarity and plain language, 123
    CTAs and other labels, 137–140
    deceptive copywriting, 134–137
    font size, 127, 143
    honest content, 134
    indecipherable, 120–123
    information clustering and hierarchy, 122, 141–142
    jargon and legalese, 124–126, 128–129

language in privacy policies (*continued*)
  navigation, 141, 142–143
  reading comprehension, 121, 131–134, 187
  scannability, 143–146
Lapse (photo-sharing app), 98–99
law enforcement
  abortion privacy searches, 6–7
  harassment of LGTBQIA+ people, 8, 199
  use of facial recognition, 9
laws of UX, 115
laws on privacy and data. *See* privacy policy, evolving impact of regulations
lawsuits, 46, 102, 121, 130, 216
learning, ongoing, 199–200, 260
Lebanon, police harassment of LGTBQIA+ people, 8
legalese in privacy policies, 124–126, 128–129
Leiser, Mark, 118
Lemonade (home insurance app), 70–71, 129–130, 144–145, 148, 219
LGTBQIA+ people
  author's work with, xx
  and Facebook change to privacy settings, 175–176
  police harassment of, 8, 199
  and real name policies, 60, 193
librarians, protection of patrons, 29
Limitless Pendant, 233
LinkedIn, 100–101, 134–135, 226
listening in, by AIs, xvi, 66, 233–235. *See also* surveillance
listening sessions, on Spotify Jam, 195–197, 200
Litman-Navarro, Kevin, 120–121
LLMs (large language models)
  ChatGPT exposure of personal data, 229
  effect on user research, 116
  training AI, 162–163
location tracking
  and Apple apps, 219–222
  Block Party extension for privacy, 245
  deceptive patterns by Google, 103–105

discoverable settings during onboarding, 156–157
  penalties and fines for, 45–46
  of Planned Parenthood patients, 7
  protection of data, 30–31
Lovett-Barron, Andrew and Ayla, 194
Luxembourg National Commission for Data Protection (CNDP), 44
Lyft, 219

M

Macy's, 10
manipulative patterns, 88
"Mapping the Temporal Dynamics of Dark Patterns" (Gray, Gairola, and Mildner), 104
The Markup, on Pixel data sharing, 6–7
Mastodon, 43
Medibank, 216
mentoring roles, 198
Messenger, 227
Meta
  AI across its platforms, 227
  backsliding on privacy and safety, 256
  Brazil action on AI and social media use, 216
  consent and transparency, 62
  fines levied, 44–45, 210
  Pixel tracker, 6–7
  platforms. *See* Facebook; Instagram; Messenger; WhatsApp
  and registration of data brokers, 214
metadata, and PII, 20
Metropolis (Lang), 235–236
Microsoft
  corporate self-regulation of privacy policies, 180, 218
  fines levied, 45, 180
  identity as a service feature, 60
  Outlook, 67–68, 113, 169
  persona spectrum, 188–189
Mildner, Thomas, 104–105, 108, 118, 155
Mims, Christopher, 4

misinformation campaigns, 236
Mismatch: How Inclusion Shapes Design (Holmes), 189, 260
MIT study, on legalese, 124
MIT Technology Review, on facial recognition, 9
Monteiro, Mike, 32, 88, 260
mosaic theory, 53–54
Mozilla
    corporate self-regulation of privacy policy, 180, 219
    Creep-O-Meter, xvi
    Firefox browser for privacy, 240–241, 253
    Firefox extension, Facebook Container, 158, 244
    Firefox onboarding, 157–158
    study on AI chatbots, 64–65
Mullvad, 240, 243
Musk, Elon, 39, 41–43, 65, 216
Myanmar genocide, 176

N

National Association of Attorneys General, xvii, 19
national data privacy laws, 215–217. *See also* General Data Protection Regulation; United States
National Literacy Institute, 131
National Public Data, 3
National Stalking Helpline, 13
Nature, study on data de-anonymizing, 232
navigation
    accessibility and inclusive design of, 187
    in privacy policies, 141, 142–143
Nazis, xvii, 26–28
"necessary" cookies, 138–139
Neidle, Dan, 123
Netherlands Organisation for Applied Scientific Research, 23, 180
New York Pass, 74
The New York Times
    on AT&T data breach, 3
    on privacy policies, 120–121, 146
    use of Tor network, 242

The New Yorker
    on Facebook change to privacy settings, 176
    use of Tor network, 242
Newsom, Gavin, 213
Next Stop Paris, 234–235
Nissan (automotive), xvi, 66
nonbinary gender titles, 55–58
Norway, fines on Grindr, 8, 44
Norwegian Data Protection Authority, 44
Nouwens, Midas, 8

O

Obama, Michelle, 38
obstruction, as deceptive pattern, 105, 108
O'Connor, Corbb, 187
onboarding, and privacy features, 67–68, 113, 156–159
onion routing, 242
online harassment, 12–13, 42, 176, 244, 246, 250
Ontario Information and Privacy Commissioner, 23
Opt-Out (Privacy Options) icon, (California), 154, 213, 257
Oremus, Will, xviii
Outlogic, 45
Ovia app, 58–59

P

passports, gender markers, 57–58
passwords, and quantum computers, 4
Patriot Act, 29
patterns. *See* deceptive patterns
PayPal, 111
Peak Design, 89–91
penalties. *See* fines and penalties
People.com, 8
Period Tracker, 59
period tracking apps, 7, 58–59, 81
Permission Slip app, 252
Perplexity (AI platform), 229
Perry, Grayson, 32, 260

INDEX   271

persona spectrums, 188–190, 193
personal information. *See also* personally identifiable information (PII)
  exercise on collecting, 117
  handling sensitively, 54–55
  in privacy definition, 17
  requesting unnecessary, and abandonment, 40–41
personalized advertising
  and deceptive language, 135–137, 139–140
  Google defaults, 150–151
  from iOS apps, 219–220
  our personal data is a product, xviii
  survey results of people's views on, 5
  targeted ads from YouTube, 159–160
personally identifiable information (PII)
  ChatGPT exposure of, 229–231
  and data deletion, 82
  defined, 20
  value of, asking only for data you need, 50–52
Pew Research
  on collection of personal information, 5
  on online harassment, 12
  on privacy policies, 120
Pfattheicher, Stefan, 109, 118
photography, as potential harm, 191, 243
Pierce, David, 233
Ping Identity, survey on abandonment, 40–41
Pixel tracker, 6–7
Placer.ai (location tracker), 7
plain language, 123, 124
Planned Parenthood, 6, 61
police. *See* law enforcement
Pope Francis, spoof of, 11
Pornhub, 48
pornography, and data deletion, 82, 85
portals for privacy features, 151–152
preferences, in privacy definition, 17
preselected checkboxes (preselection)
  adding helpful friction, 113
  as bad default, 92–93, 100–101

and deceptive patterns designed by AIs, 233
price surging, 10
privacy, defined, 15–24
  as compared with security, 18
  personal information and preferences, 17
  right to privacy, 16, 18–20, 208
  six ways to define privacy, 16
  terminology, 20–22
Privacy Badger, 244
Privacy Bee, 252
privacy by design
  check list of guidelines, 258–259
  foundational principles, 23–24, 61, 62, 181, 234
  four pillars of, 14
  future of, 256–257
  as next industry standard, 248–249
  origins of, 23
  recommended books, 260
privacy checkups, 152–153, 171–173
privacy features
  alerting users about changes, 174–176
  reminding users, 171–173
privacy features, best practices, 161–171
  agreeing and declining consent, 161
  delete everything option, 169–171
  fine-grained and overall settings, 168
  settings, private by default, 162–167, 177
  withdrawing consent, 162
privacy features, discoverability of, 154–160
  contextual placement, 154–155
  "just-in-time" alerts, 159–160
  during onboarding, 156–159
privacy features, priorities, 148–153
  privacy checkups, 152–153, 171–173
  privacy portals, 151–152
  privacy settings, 149–151, 174–176
Privacy Is Hard and Seven Other Myths (Hoepman), 53, 260
Privacy Maze, 102

Privacy Options (Opt-Out) icon, (California), 154, 213, 257
privacy policies
　as anti-patterns, 110
　language in. *See* language in privacy policies
　and Privacy Zuckering, 96–97
privacy policy, evolving impact of regulations, 207–224
　Apple's App Tracking Transparency, 219–222
　California's privacy laws, 211–214
　corporate self-regulation, 218–219
　Europe's GDPR, 208–211
　Google's cookie apocalypse, 223, 254
　other countries, 215–217
　penalty enforcement, 217–218
　U.S. state laws, 214–215. *See also* United States, state laws
privacy portals, 151–152
privacy products and tools, 239–254
　browsers, 240–242
　extensions, 244–251
　projects, 118, 253–254
　search engines, 252
　VPNs, 243–244
　websites and apps, 252
Privacy Sandbox (Google), 254
privacy settings, 149–151, 174–176
Privacy Shield agreement, 45
Privacy Zuckering, 96, 102
"Privacy Zuckering: Deceiving your Privacy by Design" (Mohit), 103
privacypatterns.org, 112
processor, as data handler, 21–22
Procter & Gamble, 213
projects for privacy, 118, 253–254
ProPublica, use of Tor network, 242
Proton Mail, 170
public by default. *See also* defaults
　in fictional book-sharing app, 204–205
　Spotify Jam, 193–195, 200
　Strava Flyby, 165–167, 177, 186
　Venmo, 93–95

Publishers Clearing House Consumer Insights, 5
purpose limitation, 49

Q

Q-Day, 4
"The Quantum Apocalypse Is Coming: Be Very Afraid" (Katwala), 4
quantum computers, and encryption technology, 4
The Quietus, 76–77
Quora, 162–163

R

radical transparency, Zuckerberg's, 62
Ray-Ban, 227
Razorfish, 13
reading, resources for learning, 118, 260
reading levels of privacy policies, 121, 131–134, 187
real name policies, 60, 193–194
RealReal (resale site), 97–98
Reasonable Expectation of Privacy Test, 19, 53
regulations. *See* privacy policy, evolving impact of regulations
reputational damage, 39–40, 217
requirements definition phase for privacy products, 251
research, inclusive, 185, 188
research phase for privacy products, 251
reversible consent, 61
right to be forgotten, 84–85, 169, 209
right to privacy, 16, 18–20, 208
"The Right to Privacy" (Warren and Brandeis), 16
risk evaluation practices, 191
Rite Aid, 10
Robertson, Andy, 165–166
robocalls, outlawed, 12
Roe v. Wade, 6–7, 34, 58
Roebert, Kevin, 244
Romantic AI, 65

RSA (network security company), 5
Ruined by Design (Monteiro), 32, 88, 260
Russia, ban of VPNs, 244

S

Sabini-Roberts, G, 55–56, 58
sabotage
    by ethical hackers in WWII, 26–28, 34
    other ways to push back, 29
Safari (browser), 157, 240
Safe Harbor agreement, 45
safety by design, 13
Safety Services Company, 137–138
Salesforce, 237
Santos, Cristiana, 108, 118
Savard Saucier, Cynthia, 100
scams, with malicious misuse of AI, 236
scannability in privacy policies, 143–146
Schlosser, Dan, 100
Schmidt, Alexandra, 198, 260
Schraffenberger, Hanna, 118
Schrems, Max, 45
Schrems I and II cases, EU, 45
Schumer, Chuck, 175
search engines, for privacy, 252
secret questions, for authentication, 110–111
security
    as compared with privacy, 18
    and data breaches, 2–3, 41
self-as-user syndrome, 32, 176, 256
#SelfCare, 133–134
Sephora (cosmetics retailer), 211
settings. *See* privacy features
Seward, Andrew, 166–167
shadow user profiles, 109
Shariat, Jonathan, 100
Shutterfly, 73
Signal, 180
sleep apps, 66
small print, 126
sneaking, 89, 105, 108
Snowden, Edward, xviii
social engineering, 108

social media
    Block Party extension for privacy, 244–250
    Brazil lawsuit on failure to warn minors, 216
    guidelines for avoiding deceptive patterns, 118
    and increase of cyberstalking and bullying, 12
    and police harassment of LGTBQIA+ people, 8
    real names and the need for anonymity, 60, 193–194
    specific platforms. *See* Facebook; Instagram; TikTok; Twitter; X; YouTube
social pyramids, 96
Social Security number, 52, 160
Solid project, 253
Solove, Daniel J., 16–17, 53, 260
Sonic internet provider, 219
South Korea, fines on companies, 45, 217
Spirit Airlines, 58
Spotify, 195–197, 200, 213
stalkerware
    accidental, 13, 165–167, 177, 186
    cyberstalking, 12–13
    in fictional book-sharing app, 204–205
Standard Contractual Clauses, 45
storage limitation, 49
Strava running app, Flyby feature, 165–167, 177, 186
subscriptions, recurring, ending, 80
"Suddenly We're Ubiquitous" (Stribley), xix
The Sun, deceptive language of, 139–140
Superbloom, 194
support networks, 198
surveillance
    by AIs, 9 10, 12, 233–235
    by Chinese, xvi, 44, 66
    in corporate policy, 219
    by government agencies, 6–7, 45
    in mosaic theory, 53
    by vacuum cleaners, cars, and sleep apps, xvi, 66

T

tables, in privacy policies, 144
"Tales from the Dark Side" (Bösch, Erb, Kargl, Kopp, and Pfattheicher), 109, 118
Targeted Speech Hearing, in headphones, 234
Tax Policy Associates, 123
"A Taxonomy of Privacy" (Solove), 53
TCL (TV manufacturer), 234–235
Terms and Conditions, 126. *See also* privacy policies
terms of use policies
  a good example, 128
  and Privacy Zuckering, 96
Tesla, xvi–xvii, 127
third parties
  as data players interacting with data, 21–22
  and data sharing, 8. *See also* data transparency
Thomson Reuters, 10
Threads, 43, 256
threat models, 191–194, 205, 243, 250
threats, in Anxiety Games, 194–195
TikTok, 44, 216
Time magazine, "Woman of the Year," 246
titles, and gender, handling responsibly on forms, 54–58
tools. *See* privacy features *entries*; privacy products and tools
Tools for Humanity, 217
top-down buddies, 198
Tor network, 242
tracking, 135. *See also* location tracking; period tracking apps
Tragic Design (Shariat and Savard Saucier), 100
transparency. *See also* data transparency
  radical (Zuckerberg), 62
Trost, Heidi, 18
Trump, Donald, 57–58, 176

trust
  erosion of, and reputational damage and abandonment, 39–40
  from honesty in content, 134
  reminding users of privacy features, 171–173
  undermined by public defaults, 167
Tumblr, 170–171
Tunisia, police harassment of LGTBQIA+ people, 8
Turkey, ban of VPNs, 244
Turow, Joseph, 124
TV, and AI-generated entertainment, 234–235
Twitter. *See also* X
  data collected from iOS apps, 219–220
  humbug headers, 137
  loss of user base, 41–43
  Musk purchase of, 42, 65
  protection of location data, 30–31
  reputational damage to, 39

U

U-Haul, 220–221
Uber, 135
Umoja Noble, Safiya, 85, 260
Understanding Privacy (Solove), 16, 260
UNESCO, human rights approach to AI, 237
United Airlines, 58
United Nations
  on Facebook and Myanmar genocide, 176
  Universal Declaration of Human Rights, 19
United States. *See also* Federal Trade Commission
  Americans with Disabilities Act (ADA), 127
  assumed consent, 62
  comprehensive federal privacy laws, 215, 257
  data privacy for EU residents, 45, 209–210
  Deceptive Experiences To Online Users Reduction (DETOUR) Act, 212

INDEX 275

United States (*continued*)
- Department of Homeland Security, 9
- destinations of fines collected, 218
- FBI, use of facial recognition, 9
- funding for Tor network, 242
- gender markers on passports, 57–58
- Immigration and Customs Enforcement (ICE), 29–30, 214
- privacy as a right, 18–19
- the right to be forgotten, 84–85
- State Data Privacy Act, 214–215

United States, state laws, 214–215
- California. *See* California
- Connecticut Data Privacy Act (CTDPA), 214–215
- Florida Digital Bill of Rights (FDBR), 214
- Illinois Biometric Information Privacy Act, 130
- Oregon, Texas, and Vermont: registration of data brokers, 214

United States Supreme Court
- Katz v. United, 19
- Reasonable Expectation of Privacy Test, 18–19, 53
- Roe v. Wade, 6–7, 34, 58
- United States v. Jones, 53

Unity, for #SelfCare, 133–134
Universal Declaration of Human Rights, 19
universities
- Carnegie Mellon, CyLab Security and Privacy Institute, opt-out icon, 213
- Indiana Univ., UX Pedagogy and Practice Lab (UXP2), study on forced options, 96
- UC Berkeley, School of Information, privacy patterns website, 112
- Univ. of Cambridge, funding for Tor network, 242
- Univ. of Michigan, School of Information, opt-out icon, 213
- Univ. of Washington, AI feature for headphones, 234

usability testing, 204, 205
USDoD (hacker), 3

user base, loss of, 41–43
user-centered design, and human-centered design, 33–34, 35
user journeys, 190
user research, LLMs' effect on, 116
user stories, 190
users
- alerting to privacy setting changes, 174–176
- designers' speaking on behalf of, 31–32, 34
- reminding of privacy features, 171–173

UX design, AI's effect on, 115–116

V

vacuum cleaners, xvi, 66
Vargas, Jose Antonio, 176
Venmo, 93–95
@venmodrugs, 95
The Verge, on Limitless Pendant, 233
Verizon, 213
Vice
- on Amazon's AI surveillance, 10
- on Placer.ai location tracking, 7

Vicemo, 94
vision, and persona spectrum, 189
visual interference, 105, 107–108
voice spoofing, and deepfakes, 11–12, 236
VPN (virtual private network), 48, 243–244

W

"walled gardens," 214, 223, 254
Walmart, 213
Warner, Mark, 212
warrant canaries, 29
Warren, Samuel, 16
Web Content Accessibility Guidelines (WCAG), 127, 187, 257
websites
- abandonment of, 40–41
- adult-oriented, 48
- for privacy, 252

WhatsApp, 96–97, 102–103, 227
"white hat" hackers, 26–28, 34, 94–95
WhosHere app, 8
Wikimedia, 125, 144
Wilson, Christo, 232
wine and cookies, 123
Wired, on third parties and data sharing, 8, 63
Woodruff, Porcha, 9
word of mouth, 39–40
working style, inclusive, 184–185
World War II, protection of identities, 26–28, 34
World Wide Web Consortium (W3C)
 Privacy Working Group, 253
 Web Accessibility Initiative (WAI), 187
 Web Content Accessibility Guidelines (WCAG), 127, 187, 257
Worldcoin Foundation, 217

X

X (formerly Twitter). *See also* Twitter
 backsliding on privacy and safety, 256
 Block Party extension for privacy, 244–245
 loss of user base, 41–43
 reputational damage to, 39
 training AI feature Grok, 42, 65, 93, 216
X-Mode, 45

Y

Yahoo (search engine), 252
YouTube
 data collected from iOS apps, 219–220
 and Duck Player, 159–160
 privacy settings, 150
YouTube History, 112

Z

Zuckerberg, Mark, 62, 96, 175–176
Zuckerman, Ethan, 62

ACKNOWLEDGMENTS

Many thanks to Lou Rosenfeld and Rosenfeld Media for believing in this project and helping me to bring it to life. Special thanks to Marta Justak for her skills and encouragement as an editor. She even told me the chapter I feared most boring was interesting. I'll let the reader decide, which chapter that was.

I'd like to thank Dr. Ann Cavoukian and others who have contributed so much to nurturing the field of privacy by design. I also owe a huge debt to many people within the global internet freedom community, especially Pepe Borrás, the founder of the Internet Freedom Festival (now Global Gathering), and others, including Gus Andrews and Georgia Bullen, who I collaborated with earlier on in my involvement in the community.

Thanks to Kyle Outlaw, my former manager and friend, for his consistent encouragement to pursue my interest in this topic during our time together at Publicis Sapient and Razorfish. He was the consummate "top-down buddy." (See Chapter 9!)

Thank you to my wife, the incredibly talented Amy Stack, for encouraging me to pursue this topic and for understanding how much of a commitment completing it would be. And for not raising an eyebrow as the pile of books I'd acquired for the sake of research grew in our Brooklyn apartment.

Thanks to my former colleagues Lorena Lima, Tim Truxell, and Tim Wolf for reviewing an early outline of this book, when it was little more than a twinkle in my eye.

A special thanks to the experts who allowed me to interview them for this book, too: Harry Brignull, Tracy Chou and Deonne Castaneda. And a huge thank you to all the technical readers who provided me with feedback based on their individual expertise, too: Pepe Borrás, Harry Brignull, Deonne Castaneda, Tal Herman, Kyle Outlaw, and Heidi Trost. This is a better book because of all of you. Thank you also to Dr. Colin Gray for reviewing specific sections of the book. And yet another thanks to Harry Brignull for writing the excellent, stage-setting foreword to this book! And my profound thanks to Jeff Atwood for his generous support for this project.

Finally, I want to sincerely thank all the third spaces where I spent an inordinate amount of time while writing this book: The Brooklyn cafés and, um, bars, as well as their amenable and friendly staff at The Center for Fiction, Cuppa Hive, The Gate, Hungry Ghost (two locations!), OS Cafe, and Postmark. Also, the branches of the Brooklyn Public Library, where I spent a lot of time researching and writing this book. Support your local library!

Dear Reader,

Thanks very much for purchasing this book. There's a story behind it and every product we create at Rosenfeld Media.

Since the early 1990s, I've been a User Experience consultant, conference presenter, workshop instructor, and author. (I'm probably best-known for having cowritten *Information Architecture for the Web and Beyond*.) In each of these roles, I've been frustrated by the missed opportunities to apply UX principles and practices.

I started Rosenfeld Media in 2005 with the goal of publishing books whose design and development showed that a publisher could practice what it preached. Since then, we've expanded into producing industry-leading conferences and workshops. In all cases, UX has helped us create better, more successful products—just as you would expect. From employing user research to drive the design of our books and conference programs, to working closely with our conference speakers on their talks, to caring deeply about customer service, we practice what we preach every day.

Please visit **rosenfeldmedia.com** to learn more about our **conferences**, **workshops**, **free communities**, and **other great resources** that we've made for you. And send your ideas, suggestions, and concerns my way: louis@rosenfeldmedia.com

I'd love to hear from you, and I hope you enjoy the book!

Lou Rosenfeld,
Publisher